Intermediate GNVQ MANUFACTURING

Jim Kelly

THOMSON
LEARNING

City and Guilds

Australia • Canada • Mexico • Singapore • Spain • United Kingdom • United States

City & Guilds Co-publishing Series

City & Guilds has a long history of providing assessments and certification to those who have undertaken education and training in a wide variety of technical subjects or occupational areas. Its business is essentially to provide an assurance that pre-determined standards have been met. That activity has grown in importance over the past few years as government and national bodies strive to create the right conditions for the steady growth of a skilled and flexible workforce.

Both teachers and learners need materials to support them as they work towards the attainment of qualifications, and City & Guilds is pleased to be working with several distinguished publishers towards meeting that need. It has been closely involved in planning, author selection and text appraisal, although the opinions expressed in the publications are those of the individual authors.

City & Guilds is fully committed to these publications and is pleased to commend them to teaching staff, students and their advisers.

Related City & Guilds/Thomson Learning titles already published
Computer-aided engineering series:
Core text: *Computer-aided Engineering*
Workbooks: *CNC Setting and Operation*; *CNC Part Programming*; *Computer-aided Draughting*; *Robot Technology*; *Programmable Logic Control*; *Drawing Standards for CAE*

To be published
GNVQ Art and Design
GNVQ Engineering
GNVQ Information Technology

Intermediate GNVQ MANUFACTURING

Jim Kelly

Australia • Canada • Mexico • Singapore • Spain • United Kingdom • United States

THOMSON

LEARNING

Intermediate GNVQ Manufacturing

Copyright © M. J. J. Kelly 1996

The Thomson Learning logo is a registered trademark used herein under licence.

For more information, contact Thomson Learning, Berkshire House, 168–173 High Holborn, London, WC1V 7AA or visit us on the World Wide Web at:
http://www.thomsonlearning.co.uk

British Library Cataloguing-in-Publication Data
A catalogue record for this book is available from the British Library

ISBN 1-86152-713-6

First published 1996 by Macmillan Press Ltd
Reprinted 2000 by Thomson Learning

Printed by Prosperous Printing Co. Ltd., China

CONTENTS

ACKNOWLEDGEMENTS

Thanks are due to those who have helped at various stages in the preparation of the text: R. Alexander, M. Amos, J. Melville, P. Hinchcliffe and A. Darbyshire.

I should like to give special thanks to my wife Irmtraud and children Michelle and Diane for their unwavering support and love.

The author and publishers thank the following for photographs:
Carlsberg-Tetley Brewing Limited (pages 21 & 168(a));
3M United Kingdom PLC Occupational Health & Environmental Safety (page 218 (b));
Merck Ltd (pages 6, 54, 77, & 168(b));
Northern Foods plc (pages 168(c) & (d), & 243)
Totectors Limited (page 218 (a), (c), (d), (e), (f), (g));
Tullis Russell *the* papermakers (pages 168(e) & 231, and the cover illustration).

Every effort has been made to trace all copyright holders, but if any have been inadvertently overlooked, the publishers will be pleased to make the necessary arrangement at the first opportunity.

INTRODUCTION

WHAT IS INTERMEDIATE GNVQ MANUFACTURING?

Intermediate GNVQ Manufacturing provides a good grounding for students who wish to progress to further study or employment where the skills, knowledge and understanding of manufacturing are useful. It is an appropriate qualification for students who wish to follow a career in any aspect of manufacturing, and for those who feel that an understanding of the sector will allow them to study and work more effectively in the future.

GNVQ Manufacturing is summed up in the following statements:

- Students following GNVQ Manufacturing programmes will gain knowledge and technical expertise in manufacturing, business and commerce. This will help them prepare for a full range of careers within the manufacturing industry, as members of a flexible, competitive, multi-skilled workforce and for progression to higher education.

- Students should develop skills and an understanding of a range of several principles on which manufacturing practices are based; and which are transferable across a broad range of products and industrial sectors including chemical, engineering, food, paper and board, steel and textiles.

- In addition to gaining knowledge of materials, production methods and processes, students will need to understand the links between such activities as identifying customer needs, design, engineering, business and commercial considerations.

WHAT DOES THIS BOOK COVER?

The book has four chapters covering the four mandatory units of the Intermediate GNVQ Manufacturing course.

Chapter 1/Unit 1	The world of manufacturing
Chapter 2/Unit 2	Working with a design brief
Chapter 3/Unit 3	Product planning, costing and quality assurance
Chapter 4/Unit 4	Manufacturing products

The four chapters cover the following areas:

- the importance of manufacturing to the UK economy;
- production systems;
- manufacturing organizations;
- environmental effects of production processes;
- designing potential products;
- producing production plans;
- calculating the cost of a product;
- preparing, processing, assembling and finishing products;
- application of quality assurance to manufacturing products.

HOW TO USE THIS BOOK

Each chapter contains a number of sections which cover the range and performance criteria of the elements within each unit. At the end of each section there is a detailed task which will enable the reader to meet the performance criteria. The chapters have been structured so that these tasks can be put together to form a report covering an element of a unit.

Let me give you an example:

EXAMPLE

Unit 1 comprises four elements and Chapter 1 is structured to provide four reports covering the evidence indicators for each of these four elements. Let's look more closely at element 2 of Unit 1, called 'Investigate production systems'. There are five performance criteria for the student to meet to complete this element. Correspondingly, within Chapter 1, sections 4–8 inclusive cover the range, and tasks 8–12 inclusive provide the reader with the means to meet the performance criteria of element 2 of Unit 1.

(part of) CHAPTER 1			
Section	Title	Task	Evidence
1.4	Key production stages	8	Brief report and flow diagram
1.5	Scales of production	9	Brief report
1.6	Production systems	10	Flow diagram
1.7	Quality control	11	Flow diagram
1.8	Effects of changing the scales of production	12	Table
		8–12	Report titled: 'Investigate production systems' using the above evidence

Tasks 8–12 are all cross-referenced so that readers know that they come together to form one report titled 'Investigate production systems' covering element 2 of Unit 1.

All the other elements are covered in a similar manner, and are shown in the Appendix.

In Chapter 2, two product proposals are originated and presented, and these two products are used throughout Chapters 3 and 4.

WHAT ABOUT CORE SKILLS?

All Intermediate students have to gain core skill units in Application of Number, Communication and Information Technology at level 2. This book provides the reader with plenty of opportunities to tackle real problems and issues in vocational settings. Within each task the reader should identify what skills have been used and where evidence might be used for gaining credit towards the units.

1 · THE WORLD OF MANUFACTURING

1.1 WHAT IS MANUFACTURING?

Manufacturing is the making of goods by physical labour and machinery. It generally involves the conversion of raw materials into a finished product. Raw materials include such items as fruit, cloth, wood, and plastic and they are supplied from industries such as agriculture and mining. Finished products include books, jeans and compact discs and are supplied to customers.

When raw materials are converted into a finished product there is VALUE ADDED to them. This increase in value can greatly benefit the economy. Manufacturing makes a vital contribution to both local and national economies of the United Kingdom (UK). Many areas in the UK rely heavily on manufacturing industries for both employment and local economic survival. Manufactured goods account for some 80% of the UK exports and 20% of the Gross Domestic Product (GDP). The UK manufacturing industry employs 4.3 million people and that accounts for 20.2% of all those in employment (*December 1993 figures*).

> **DEFINITION**
> **Gross Domestic Product** (GDP): the total value of all goods and services produced by a nation in a year.

1.2 MAIN UK MANUFACTURING SECTORS

> MAIN UK
> MANUFACTURING SECTORS
> * chemical
> * engineering
> * food, drink and tobacco
> * paper and board
> * printing and publishing
> * textiles, clothing and footwear

Let's begin by investigating the chemical industry.

➤ Chemical industry

The chemical industry is divided into quite a few different subsectors, and is classified as a medium technology industry. Organic and inorganic chemical production involves the conversion of feedstock (raw materials) into base chemicals which are used extensively throughout the other chemical industry subsectors in the manufacture of their products.

Both organic and inorganic chemicals are produced in very high tonnages and hence the chemical plants are large and tend to be well established. They are highly complex installations and represent a major capital investment, and thus production is dominated by just a few chemical companies. These companies are also involved in using these chemicals to manufacture such products as plastics and paints.

Within the industry, there are a number of multi-plant sites where a variety of products from the different subsectors are manufactured.

The chemical industry depends upon technological development for its growth and consequently research and development departments are found in all major companies.

Below you will find a table of the top 20 chemical companies in the UK listed in order of their turnover for the period ending 1992 or 1993.

No	UK company	Turnover (£ millions)	Number of employees
1	Imperial Chemical Industries PLC	10 632	87 100
2	BOC Group PLC (The)	3 068	38 434
3	Courtaulds PLC	1 995	19 500
4	Harrisons & Crosfield PLC	1 933	32 177
5	Laporte PLC	877	7 330
6	British Vita PLC	754	12 808
7	3M UK Holdings PLC	542	5 021
8	Exxon Chemical Ltd	437	1 747
9	Croda International PLC	415	4 378
10	Ellis & Everard PLC	395	1 862
11	Hickson International PLC	368	2 840
12	Dow Chemical Co Ltd	300	669
13	Allied Colloids Group PLC	295	2 796
14	Yule Catto & Co PLC	222	2 799
15	Air Products PLC	218	2 378
16	Rohm and Haas (UK) Ltd	217	549
17	British Polythene Industries PLC	212	2 514
18	BTP PLC	209	2 227
19	Sterling-Winthrop Group Ltd	197	2 388
20	Grace (WR) Ltd	180	2 017

Top 20 UK chemical companies
Source: *The Times 1000*, 1995 (Times Books)

The UK chemical industry is very successful and is one of the largest in Europe. Imperial Chemical Industries PLC is the largest chemical group in the UK and accounts for over one quarter of the total UK production. Large-sized companies such as BOC Group PLC and medium-sized companies such as the Dow Chemical Co Ltd account for a further half of the total production.

And on the opposite page is a table of the top 20 health and household companies in the UK in order of their turnover for the period ending 1992 or 1993.

Production plant for liquid crystals

No	UK Company	Turnover (£ millions)	Number of employees
1	Smith Kline Beecham PLC	6 164	52 700
2	Glaxo Holdings PLC	4 930	40 024
3	ZENECA Group PLC	4 440	32 300
4	Reckitt & Colman PLC	2 096	21 000
5	Wellcome PLC	2 041	17 571
6	AAH PLC	1 402	8 471
7	Fisons PLC	1 324	13 364
8	Procter & Gamble Ltd	1 255	5 610
9	Unichem Group PLC	1 178	4 865
10	Smith & Nephew PLC	949	13 099
11	MFI Furniture Group PLC	604	7 579
12	Rentokil Group PLC	588	22 756
13	Eclipse Blinds PLC	424	3 434
14	London International Group PLC	416	10 250
15	Sony Music Entertainment (UK) Group Ltd	369	1 554
16	Bristol-Myers Squibb Holdings Ltd	324	2 584
17	Amersham International PLC	269	3 261
18	Merck Sharp & Dohme (Holdings) Ltd	251	2 480
19	Paterson Zochonis PLC	233	3 926
20	Lilly [Eli] Group Ltd	224	2 102

Top 20 UK health and household companies
Source: *The Times 1000*, 1995 (Times Books)

Let's take a closer look at the organic and inorganic chemical production and see what typical products are manufactured with the base chemicals by the other subsectors.

◆ Organic chemicals

Organic chemicals form the major part of the petrochemical industry and are produced from petroleum-based chemicals such as naphtha. Organic chemicals such as ethylene, propylene and benzene are used in the manufacture of pharmaceuticals, solvents, plastics, detergents, synthetic rubber and resins, pesticides, and fibres.

Here is a table of organic chemicals showing the raw material feedstock from which they are processed, and the typical products for which they are used in manufacture.

Raw material (feedstock)	Organic chemical	Typical products
Natural gas	Methanol	Plastics, solvents
Hydrocarbon	Olefins including ethylene, propylene, butylene	Plastics, solvents, synthetic rubber
Olefin plants and naphtha	Aromatics including benzene, toluene, xylene	Plastics, detergents, pesticides, pharmaceuticals, explosives, dyestuffs, fibres

Organic chemicals

◆ Inorganic chemicals

Inorganic chemicals are produced from raw materials such as ammonia and sulphur and include nitric and sulphuric acid. Inorganic chemicals are used in the manufacture of fertilisers, glass, paint, pigments, explosives, and photographic materials.

Below you will see a table of inorganic chemicals showing the raw material feedstock from which they are processed, and the typical products for which they are used in manufacture.

Raw material (feedstock)	Inorganic chemical	Typical products
Natural gas	Ammonia	Fertilisers, plastics, nylon
Ammonia	Nitric acid	Fertilisers, pharmaceuticals, explosives, dyestuffs
Sulphur	Sulphuric acid	Fertilisers, plastics, detergents, paints, pigments, dyestuffs
Ammonia & brine	Sodium carbonate (soda ash)	Glass, soap, detergents

Inorganic chemicals

Here is a list of the main products manufactured by the other chemical industry subsectors:

- fertilisers;
- paint and varnish;
- pharmaceuticals including antibiotics and naturally occurring drugs;
- detergents including soap, detergents, perfumes, cosmetics, toilet preparations;
- dyestuffs and pigments;
- polymers including plastics, rubbers and resins, and miscellaneous products including photographic materials, formulated adhesives, furniture polishes, explosives, printing ink.

Now let's have a look at the import and export trade of the UK chemical industry.

CHEMICAL				
Year	Import trade (£ millions)	% of total UK import trade	Export trade (£ millions)	% of total UK export trade
1992	11 618	9.2	14 974	13.8

Import and export trade of the UK chemical industry

And what about the number of employees working in the UK chemical industry?

CHEMICAL			
Period	No of employees	% of all manufacturing industry employees	% of all those in employment
Dec 1993	300 100	7.1	1.4

Number of employees in the UK chemical industry

This task is carried out in conjunction with tasks 2, 3, 4, 5, 6, and 7

TASK 1

An investigation into the UK chemical manufacturing sector

To carry out this investigation, use all the resources available to you – such as the library, magazines, journals, teachers, tutors, local chemical companies, and 'yellow pages'.

- Into what subsectors is the UK chemical manufacturing sector divided?
- What is the typical range of products manufactured by the UK chemical manufacturing sector?
- Who are the main companies in the UK chemical manufacturing sector?
- Is the UK chemical manufacturing sector dominated by just a few large companies or are there lots of small and medium-sized companies?
- Where are the main chemical companies located regionally in the UK?
- How many workers are employed in the UK chemical manufacturing sector?
- What are the import and export trade figures for the UK chemical manufacturing sector?

Produce a brief report on the UK chemical manufacturing sector using the above questions as topic headings.

Keep the brief report in your portfolio.

- metal manufacture
- metal products
- mechanical engineering
- electrical and electronic engineering
- instrument engineering
- transport manufacture

Now let's consider the engineering industry:

➤ Engineering industry

It is a very diverse industry, divided into a number of different subsectors ranging from the relatively low technology of fabricating metal products to the high technology aerospace industry.

Let's begin by looking at the metal manufacturing industry:

◆ Metal manufacture

The metal manufacturing sector is dominated by the iron and steel industry. The non-ferrous metal manufacturing industry is one of the largest in Europe producing aluminium, copper, lead, zinc, nickel and tin.

The UK is also a major producer of the following metals:

uranium
beryllium
zirconium

silicon
germanium
selenium

niobium

titanium

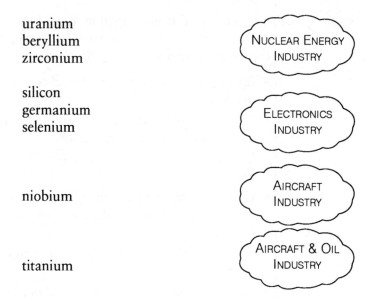

NUCLEAR ENERGY INDUSTRY

ELECTRONICS INDUSTRY

AIRCRAFT INDUSTRY

AIRCRAFT & OIL INDUSTRY

The finished products of the metal manufacturing industry are sold for subsequent fabrication in other industries such as other engineering subsectors, and construction.

British Steel PLC is by far the largest company, responsible for about 80% of the UK steel production. The main steel-producing areas are Yorkshire, Humberside and Wales. The non-ferrous metal industry is centred in the Midlands, with some operations on Tyneside, London, Avonmouth and South Wales.

Let's look at the chart showing the top 20 metal manufacturing and metal-forming companies in the UK in order of their turnover for the period ending 1992 or 1993.

No	UK Company	Turnover (£ millions)	Number of employees
1	British Steel PLC	4 303	45 600
2	Engelhard Ltd	3 388	419
3	Johnson Matthey	1 854	6 223
4	Glynwed International PLC	966	11 149
5	British Alcan Aluminium PLC	748	7 718
6	Inco Europe Ltd	601	2 978
7	ASW Holdings PLC	429	2 732
8	Bridon PLC	316	4 464
9	Pasminco Europe Ltd	198	848
10	Murray International Holdings Ltd	165	1 825
11	Co-Steel (UK) Ltd	164	1 088
12	Mount Isa Holdings (UK) Ltd	163	508
13	Metalchem International Ltd	140	41
14	Johnson & Firth Brown PLC	130	2 287
15	Sheffield Forgemasters Ltd	123	2 021
16	Lazarus (Leopold) Ltd	105	35
17	Avdel PLC	102	1 789
18	Cook [William] PLC	101	2 073
19	Coopers Holdings Ltd	100	410
20	Royal Mint Trading Fund	83	1 040

Top 20 UK metal manufacturing and metal forming companies Source: *The Times 1000*, 1995 (Times Books)

Now let's investigate the manufacture of metal products:

◆ Metal products

The metal product manufacturing industry is made up of a very large number of engineering firms. Typical products manufactured are:

- hand tools including files, hammers, saws, spades;
- cutlery and tableware;
- bolts, nuts, nails, screws, needles, pins, locks and keys;
- wire;
- cans and metal boxes, metal furniture and windows;
- jewellery and precious metal goods such as gold and silverware;
- small tools and gauges including jigs and fixtures, press tools and moulds, hard metal-tipped tools, metal-cutting tools.

In metal product manufacturing, there are many small firms – but less than a dozen large firms produce more than half of the total output.

Now let's look at the mechanical engineering industry:

◆ Mechanical engineering

This consists of a group of industries, including those who manufacture:

- machine tools;
- plant and machinery;
- construction equipment;
- mechanical handling equipment;
- agricultural equipment;
- office machinery;
- general machinery;
- miscellaneous machinery.

Let's have a look at each of these manufacturing industries:

◆ Machine tool manufacturing

This includes the manufacture of the following:
- milling, grinding, and turning machines;
- numerically controlled machine tools including machine centres.

◆ Plant and machinery manufacturing

This involves the manufacture of:
- metallurgical furnaces;
- lime and cement kilns;
- water and sewage treatment plant;
- fabricated steelwork for bridges;
- buildings and industrial installations;
- machines for food and drink preparation, processing and sterilisation.

◆ Construction manufacturing

This consists of the manufacture of the following equipment:
- excavating;
- earth-moving;
- road-making;
- piledriving;
- quarry crushing.

◆ Mechanical handling equipment manufacturing

This includes the manufacture of:
- cranes;
- bridge transporters;
- excavators;
- lifts;
- elevators;
- hoists;
- conveyors.

◆ Agriculture machinery manufacturing

This consists of the manufacture of many special purpose machines such as:
- root harvesters;
- fruit harvesters.

◆ Office machinery manufacturing

This includes the manufacture of:
- photocopiers;
- mail-room equipment;
- microfilm readers;
- accounting machines.

◆ General mechanical engineering manufacturing

This industry supplies parts and components and undertakes general sub-contracting, fabricating and repair work.
   Some important manufacturing includes:
- various types of bearing;
- chains;
- drop-forgings.
- fasteners;
- gears;

◆ Miscellaneous machinery manufacturing

This includes:
- space heating equipment;
- ventilating and air-conditioning equipment;
- hand tools;
- automatic vending machines;
- other types of specialised equipment.
- garden tools;
- packaging and bottling machinery;

Now let's look at the top 20 general engineering companies in the UK in order of their turnover for the period ending 1992 or 1993.

No	UK Company	Turnover (£ millions)	Number of employees
1	BICC PLC	3 614	39 151
2	Lucas Industries PLC	2 434	48 866
3	AMEC PLC	2 184	25 681
4	Tomkins PLC	2 060	30 511
5	GKN PLC	2 022	27 487
6	T & N PLC	1 662	41 220
7	BBA Group PLC	1 417	21 388
8	TI Group PLC	1 393	24 000
9	IMI PLC	1 065	17 196
10	APV PLC	906	11 318
11	FKI PLC	756	12 470
12	Babcock International Group PLC	748	13 544
13	Weir Group PLC (The)	449	6 245
14	AEA Technology	415	9 195
15	Senior Engineering Group PLC	390	5 350
16	Simon Engineering PLC	386	5 585
17	Cummins UK Ltd	368	4 163
18	Staveley Industries PLC	345	5 303
19	Howden Group PLC	335	4 226
20	McKechnie PLC	314	5 763

Top 20 UK general engineering companies Source: *The Times 1000*, 1995 (Times Books)

Right, let's investigate the next engineering subsector:

◆ Electrical and electronic engineering

The electrical engineering industry manufactures the following types of product:

- generators;
- motors;
- transformers;
- switchgear;
- control gear;
- optical fibres.
- turbines;
- converters;
- rectifiers;
- starting gear;
- cables;

The electrical goods market is dominated by a few large companies who manufacture the following products:

- refrigerators;
- irons;
- washing machines;
- heaters;
- freezers;
- kettles;
- dryers;
- cookers.

The electrical engineering industry also manufactures electrical equipment for the transport industry, and also electrical light fittings and lamps.

And here are the top 20 electrical and electronic engineering companies in the UK in order of their turnover for the period ending 1992 or 1993:

No	UK Company	Sector	Turnover (£ millions)	Number of employees
1	General Electric Co PLC	Electronics	5 612	93 228
2	IBM UK Holdings Ltd	Electronics	3 751	19 651
3	Rank Xerox Ltd	Electronics	3 161	24 422
4	ICL PLC	Electronics	2 478	13 973
5	Siebe PLC	Electronics	1 619	29 644
6	Sony United Kingdom Ltd	Electronics	1 190	3 709
7	Hewlett-Packard Ltd	Electronics	1 025	4 054
8	Gestetner Holdings PLC	Electronics	1 000	11 007
9	Delta PLC	Electrical	833	13 521
10	Digital Equipment Co Ltd	Electronics	800	6 166
11	Chubb Security PLC	Electronics	674	19 360
12	Panasonic UK Ltd	Electrical	640	765
13	Hitachi Europe Ltd	Electronics	637	621
14	Digital Equipment Scotland Ltd	Electronics	540	1 561
15	Marlowe Holdings Ltd	Electrical	468	2 944
16	Black & Decker International	Electrical	442	6 177
17	CEF Holdings Ltd	Electrical	410	3 930
18	Electrocomponents PLC	Electronics	388	2 757
19	TT Group PLC	Electronics	358	6 843
20	Mitsubishi Electric UK Ltd	Electronics	348	2 171

Top 20 UK electrical and electronic engineering companies Source: *The Times 1000, 1995* (Times Books)

The electronics engineering industry is one of the largest in the world and manufactures the following types of equipment:

- telecommunications equipment;
- computer equipment;
- radar equipment;
- navigational aids;
- security systems;
- process control equipment;
- consumer goods such as music centres, hi-fi equipment, compact disc players, radios, televisions.

The electronics industry also manufactures a wide range of electronic components, which includes the following:

- transistors;
- diodes;
- resistors;
- capacitors;
- inductors;
- integrated circuits (ICs).

The electrical and electronics industry is dominated by General Electric Company (GEC) PLC who are located throughout the UK. GEC are Europe's biggest avionics company, and the UK's largest employer in the space business. The company also employs more people in engineering in Scotland than any other private sector firm.

Now let's look at the next engineering subsector:

◆ Instrument engineering

Examples of the manufacture of scientific and industrial instruments:

- process measurement and control instruments;
- optical instruments;
- electronic measurement and test instruments;
- automatic test equipment;
- non-destructive test equipment;
- medical diagnosis equipment;
- pollution control equipment.

The manufacture of **photographic equipment** includes:

- photographic cameras;
- cinematographic cameras;
- projectors.

Other products manufactured in the instrument industry are:

- watches;
- clocks;
- surgical instruments and appliances.

Here are the top six instrument engineering companies in the UK in order of their turnover for the period ending 1992 or 1993.

Top six UK instrument engineering companies
Source: *The Times 1000*, 1995 (Times Books)

No	UK Company	Turnover (£ millions)	Number of employees
1	Kodak Ltd	1 003	8 384
2	Canon (UK) Ltd	395	2 533
3	Schlumberger PLC	315	5 381
4	Life Sciences International PLC	154	1 893
5	Halma PLC	116	1 902
6	Graseby PLC	114	1 503

Let's consider the last sector, the transport manufacturing industry:

◆ Transport manufacture

The transport manufacturing industry consists of the following subsectors:

- motor vehicles
- motor cycles
- railway equipment;
- shipbuilding;
- tractors;
- pedal cycles;
- aerospace engineering;
- marine engineering.

We'll take them one by one. But first, here is a chart listing the top 20 transport manufacturing companies in the UK in order of their turnover for the period ending 1992 or 1993.

No	UK Company	Turnover (£ millions)	Number of employees
1	British Aerospace PLC (Aerospace)	10 760	96 800
2	Inchcape PLC	5 877	38 189
3	Ford Motor Co Ltd	5 383	38 400
4	Lex Service PLC	5 095	61 113
5	Rolls Royce PLC (Aerospace)	3 518	49 200
6	Vauxhall Motors Ltd	2 998	11 042
7	Honda Motor Europe Ltd	1 069	1 225
8	Nissan Motor Manufacturing (UK) Ltd	1 054	4 974
9	Cowie Group PLC	800	3 676
10	Smiths Industries PLC (Aerospace)	723	11 535
11	Vickers PLC (Aerospace)	690	9 406
12	BSG International PLC	574	6 112
13	IBC Vehicles Ltd	558	2 432
14	Automative Financial Group Holdings Ltd	546	3 735
15	Hartwell PLC	454	2 685
16	Evans Halshaw Holdings PLC	405	1 910
17	Short Brothers PLC (Aerospace)	392	8 982
18	Henlys Group PLC	372	2 559
19	Appleyard Group PLC	367	1 915
20	Lookers PLC	363	2 163

Top 20 UK transport manufacturing companies Source: *The Times 1000*, 1995 (Times Books)

Let's start with the motor vehicle industry.

◆ Motor vehicle industry

This manufactures the following:
- cars;
- caravans;
- parts and components;
- buses;
- vans;
- trailers;
- commercial goods vehicles;
- coaches.

◆ Tractor industry

The UK is a major world producer and exporter of tractors.

◆ Motor cycle & pedal cycle industry

The UK motor cycle industry has contracted although the pedal cycle industry is expanding.

◆ Railway equipment manufacturing industry

This provides:
- locomotives;
- traction and control gear;
- track equipment;
- rolling stock;
- signalling;
- electrification of rail tracks.

◆ UK aerospace industry

This manufactures the following types of product:
- aircraft (civil and military);
- guided weapons;
- space vehicles;
- helicopters;
- hovercraft;
- aero-engines.

British Aerospace PLC is the UK's largest manufacturer of engineering-based products, and one of the world's leading aerospace organizations. Short Brothers PLC is the largest engineering company in Northern Ireland, employing almost 10 000 people in the manufacture of commercial and military aircraft, aerostructures and defence systems. Rolls Royce is a major manufacturer of aero-engines for the aerospace industry.

◆ Aviation equipment manufacturing industry

This provides:
- electrical, mechanical and hydraulic power systems;
- engine and flight controls;
- cabin furnishings;
- radar and air traffic control equipment.
- flight deck information displays;
- flight simulators;

Let's have a look at the import and export trade of the UK engineering industry starting with iron and steel:

IRON AND STEEL				
Year	Import trade (£ millions)	% of total UK import trade	Export trade (£ millions)	% of total UK export trade
1992	2 514	2.0	3 005	2.8

NON-FERROUS METALS				
Year	Import trade (£ millions)	% of total UK import trade	Export trade (£ millions)	% of total UK export trade
1992	2 590	2.1	1 753	1.6

What about the number of employees working in the UK engineering industry?

Year	Sector	No of employees	% of all manufacturing employees	% of all those in employment
Dec 1993	metal manufacture	116 800	2.7	0.6
Dec 1993	mechanical engineering	592 600	13.9	2.8
Dec 1993	electrical and electronic engineering	453 200	10.6	2.2
Dec 1993	instrument engineering	83 700	2.0	0.4
Dec 1993	transport manufacture	373 000	8.8	1.8
Dec 1993	metal products	2 088 000	49.1	9.9
Dec 1993	TOTAL	3 707 300	87.1	17.7

This task is carried out in conjunction with tasks 1, 3, 4, 5, 6, and 7

TASK 2

An investigation into the UK engineering manufacturing sector

To carry out this investigation, use all the resources available to you – such as the library, magazines, journals, teachers, tutors, local engineering companies, and 'yellow pages'.

- Into what subsectors is the UK engineering manufacturing sector divided?
- What is the typical range of products manufactured by the UK engineering manufacturing sector?
- Who are the main companies in the UK engineering manufacturing sector?
- Is the UK engineering manufacturing sector dominated by just a few large companies or are there lots of small and medium-sized companies?
- Where are the main engineering companies located regionally in the UK?
- How many workers are employed in the UK engineering manufacturing sector?
- What are the import and export trade figures for the UK engineering manufacturing sector?

Produce a brief report on the UK engineering manufacturing sector using the above questions as topic headings.

Keep the brief report in your portfolio.

FOOD, DRINK & TOBACCO
INDUSTRY SUBSECTORS

- dairy industry
- processed foods industry
- bakery products
- confectionery industry
- oils and fats
- miscellaneous food products
- drinks industry
- tobacco industry

And now let's investigate the food, drink and tobacco industry.

➤ Food, drink and tobacco industry

The UK is one of the world's leading manufacturers of food and drink products. The industry as a whole comprises eight subsectors:

Here is a table of the top 20 food, drink and tobacco companies in the UK in order of their turnover for the period ending 1992 or 1993.

No	UK Company	Sector	Turnover (£ millions)	Number of employees
1	BAT Industries PLC	Tobacco	17 879	88 358
2	Unilever PLC	Food	8 440	N/A
3	Grand Metropolitan PLC	Drink	8 120	69 940
4	Allied-Lyons PLC	Drink	5 526	55 385
5	Gallaher Ltd	Tobacco	4 712	24 835
6	Guinness PLC	Drink	4 663	23 275
7	Hillsdown Holdings PLC	Food	4 595	43 251
8	Dalgety PLC	Food	4 470	15 417
9	Bass PLC	Drink	4 451	81 105
10	Associated British Foods PLC	Food	4 422	50 659
11	Cadbury Schweppes PLC	Food	3 724	39 066
12	Tate & Lyle PLC	Food	3 698	15 800
13	United Biscuits (Holdings) PLC	Food	3 049	39 352
14	Whitbread PLC	Drink	2 360	62 389
15	Northern Foods PLC	Food	2 026	30 219
16	Unigate PLC	Food	1 925	25 385
17	Scottish & Newcastle PLC	Drink	1 514	28 487
18	Hoops Ltd	Food	1 430	6 009
19	Cowage Ltd	Drink	1 333	6 618
20	Western United Investment Co Ltd	Food	1 312	18 962

Top 20 UK food, drink and tobacco companies Source: *The Times 1000, 1995* (Times Books)

◆ The food industry

◆ The dairy industry

Let's begin with this industry: the milk sold by farmers is heat treated by dairies for liquid consumption, or manufactured into milk products such as cheese, butter, cream, yoghurt, evaporated and condensed milk.

◆ The processed foods industry

Processed foods can be split into two main sectors, ie frozen fruit and vegetables, and meat and fish.

The frozen fruit and vegetables industry

This includes:
- canned fruit;
- potato chips;
- pickles;
- canned vegetables;
- crisps;
- sauces.

The meat and fish industry

This includes:
- cooked meat;
- canned meat;
- smoked fish;
- smoked meat;
- cooked fish;
- canned fish.

◆ The bakery industry

Bakery products include:
- bread;
- cakes;
- pastries;
- puddings.
- biscuits;
- grain, flours, meal;
- pies;

The bakery products industry is dominated by Allied Bakeries and Rank Hovis McDougal. Smaller bakeries produce bread and cakes.

◆ The confectionery industry

Confectionery products include:
- chocolate;
- sugar confectionery.
- cocoa;

The confectionery industry is made up of many small and medium-sized companies. The three largest confectionery product manufacturers are Mars, Cadbury Schweppes and Rowntree Mackintosh.

◆ The oils and fats industries

These include the manufacture of the following types of product:
- vegetable oil;
- white fats;
- cooking oil;
- margarine.

◆ Miscellaneous foods

Miscellaneous food products that are manufactured include:
- starch products;
- salt;
- tea;
- dried powder products;
- pet food.
- glucose products;
- coffee;
- yeast products;
- animal food;

◆ The drinks industry

The drinks industry is divided into:

- spirits distilling;
- wines;
- brewing;
- soft drinks.

◆ Spirits distilling industry

This is dominated by the production of whisky, particularly Scotch whisky. Other spirits which are distilled include gin, vodka and rum.

◆ Brewing industry

Most of the beer production in the UK is provided by less than ten major brewers, with about 80 smaller brewers producing the traditional and regional beers.

◆ Wine industry

There is a small but expanding wine industry in the UK, with most grapes imported and the remainder grown in vineyards centred in southern England.

Ciders and perrys are produced by three main groups located in Devon, Somerset, and Hereford and Worcester.

◆ Soft drinks industry

Beer production

This sector of the drinks industry is large and still expanding; it manufactures various types of drink, such as:
- carbonated;
- concentrates;
- cordials;
- squash;
- fruit juices;
- tonic waters.

◆ The tobacco industry

The tobacco industry cures imported leaf, and then cuts and prepares it for the manufacture of:

- cigars;
- rolling tobacco;
- pipe tobacco;
- cigarettes.

The UK tobacco industry is made up of four major companies, the largest being BAT Industries PLC, together with a number of smaller companies.

Now let's have a look at the import and export trade of the UK food, drink and tobacco industry starting with fruit and vegetables.

FRUIT AND VEGETABLES				
Year	Import trade (£ millions)	% of total UK import trade	Export trade (£ millions)	% of total UK export trade
1992	3 118	2.5	330	0.3

COFFEE, TEA, COCOA, AND SPICES				
Year	Import trade (£ millions)	% of total UK import trade	Export trade (£ millions)	% of total UK export trade
1992	891	0.7	509	0.5

CEREAL AND CEREAL PREPARATIONS				
Year	Import trade (£ millions)	% of total UK import trade	Export trade (£ millions)	% of total UK export trade
1992	1 004	0.8	1 204	1.1

MEAT AND MEAT PREPARATIONS				
Year	Import trade (£ millions)	% of total UK import trade	Export trade (£ millions)	% of total UK export trade
1992	2 033	1.6	827	0.8

DRINK AND TOBACCO				
Year	Import trade (£ millions)	% of total UK import trade	Export trade (£ millions)	% of total UK export trade
1992	2 026	1.6	3 417	3.2

ANIMAL AND VEGETABLE OILS AND FATS				
Year	Import trade (£ millions)	% of total UK import trade	Export trade (£ millions)	% of total UK export trade
1992	423	0.3	83	0.1

And what about the number of employees working in the UK food, drink and tobacco industry?

FOOD, DRINK AND TOBACCO				
Year	Sector	No of employees	% of all manufacturing employees	% of all those in employment
Dec 1993	food	419 000	9.8	2.0
Dec 1993	drink and tobacco	72 100	1.7	0.3

This task is carried out in conjunction with tasks 1, 2, 4, 5, 6, and 7

TASK 3

An investigation into the UK food, drink and tobacco manufacturing sector

To carry out this investigation, use all the resources available to you – such as the library, magazines, journals, teachers, tutors, local food, drink and tobacco companies, and 'yellow pages'.

- Into what subsectors is the UK food, drink and tobacco manufacturing sector divided?
- What is the typical range of products manufactured by the UK food, drink and tobacco manufacturing sector?
- Who are the main companies in the UK food, drink and tobacco manufacturing sector?
- Is the UK food, drink and tobacco manufacturing sector dominated by just a few large companies or are there lots of small and medium-sized companies?
- Where are the main food, drink and tobacco companies located regionally in the UK?
- How many workers are employed in the UK food, drink and tobacco manufacturing sector?
- What are the import and export trade figures for the UK food, drink and tobacco manufacturing sector?

Produce a brief report on the UK food, drink and tobacco manufacturing sector using the above questions as topic headings.

Keep the brief report in your portfolio.

Now let's take a look at the paper and board manufacturing industry.

➤ Paper and board industry

The UK paper and board industry manufactures paper and board from pulp. It is one of the largest in Europe and produces a wide range of different products – from household tissues to cardboard boxes.

Let's take a look at a table of the top 20 packaging, paper and printing companies in the UK in order of their turnover for the period ending 1992 or 1993.

No	UK Company	Turnover (£ millions)	Number of employees
1	Arjo Wiggins Appleton PLC	2 727	18 771
2	Bowater PLC	2 112	26 400
3	Bunzi PLC	1 520	8 046
4	De La Rue PLC	593	7 743
5	Linpac Group Ltd	532	7 075
6	Smith [David S] (Holdings) PLC	519	6 304
7	Wace Group PLC	336	4 821
8	Fine Art Developments PLC	312	4 613
9	Scott Paper (UK) Ltd	310	2 414
10	BPCC Ltd	290	5 310
11	Kimberly-Clark Ltd	264	2 950
12	UK Paper PLC	229	1 925
13	Waddington [John] PLC	222	3 486
14	St Ives PLC	221	3 027
15	Portals Group PLC	198	2 892
16	Reader's Digest Association Ltd (The)	192	946
17	Sidlaw Group PLC	173	1 980
18	Watmoughs (Holdings) PLC	150	2 052
19	Field Group PLC	139	1 895
20	Donnelley [RR] Ltd	132	943

Top 20 UK packaging, paper and printing companies
Source: *The Times 1000*, 1995 (Times Books)

The UK paper and board industry manufactures the following types of product:

- newsprint;
- printing and writing paper for books, magazines, stationery;
- wrapping and packing paper for wrappings, bags, sacks;
- household tissues including toilet paper, paper tissues and hand-kerchiefs;
- packaging boards.

There is also a paper and board conversion industry, and this manufactures the following range of products:

- stationery including binders and notepaper;
- packaging products including boxes and bags;
- wallpaper.

Let's have a look at the import and export trade of the UK paper, board and pulp industry.

PAPER, BOARD AND PULP				
Year	Import trade (£ millions)	% of total UK import trade	Export trade (£ millions)	% of total UK export trade
1992	3 801	3	1 730	1.6

And what about the number of employees working in the UK paper, board and pulp industry?

PAPER, BOARD AND PULP				
Period	No of employees	% of all manufacturing industry employees	% of all those in employment
Dec 1993	113 500	2.7	0.5

This task is carried out in conjunction with tasks 1, 2, 3, 5, 6, and 7

TASK 4

An investigation into the UK paper and board and manufacturing sector

To carry out this investigation, use all the resources available to you – such as the library, magazines, journals, teachers, tutors, local paper and board companies, and 'yellow pages'.

- Into what subsectors is the UK paper and board manufacturing sector divided?
- What is the typical range of products manufactured by the UK paper and board manufacturing sector?
- Who are the main companies in the UK paper and board manufacturing sector?
- Is the UK paper and board manufacturing sector dominated by just a few large companies or are there lots of small and medium-sized companies?
- Where are the main paper and board companies located regionally in the UK?
- How many workers are employed in the UK paper and board manufacturing sector?
- What are the import and export trade figures for the UK paper and board manufacturing sector?

Produce a brief report on the UK paper and board manufacturing sector using the above questions as topic headings.

Keep the brief report in your portfolio.

Right, now let's investigate the printing and publishing industry.

➤ Printing and publishing industry

Mergers in the newspaper, magazine and book publishing sectors have led to the industry having a few groups controlling it. But there are a large number of small firms involved in bookbinding and general printing.

The top 20 UK companies in packaging, paper and printing were shown in the last section.

Let's have a look at the number of employees in the UK printing and publishing industry:

PRINTING AND PUBLISHING			
Period	No of employees	% of all manufacturing industry employees	% of all those in employment
Dec 1993	336 400	7.9	1.6

This task is carried out in conjunction with tasks 1, 2, 3, 4, 6, and 7

TASK 5

An investigation into the UK printing and publishing manufacturing sector

To carry out this investigation, use all the resources available to you – such as the library, magazines, journals, teachers, tutors, local printing and publishing companies, and 'yellow pages'.

- Into what subsectors is the UK printing and publishing manufacturing sector divided?
- What is the typical range of products manufactured by the UK printing and publishing manufacturing sector?
- Who are the main companies in the UK printing and publishing manufacturing sector?
- Is the UK printing and publishing manufacturing sector dominated by just a few large companies or are there lots of small and medium-sized companies?
- Where are the main printing and publishing companies located regionally in the UK?
- How many workers are employed in the UK printing and publishing manufacturing sector?
- What are the import and export trade figures for the UK printing and publishing manufacturing sector?

Produce a brief report on the UK printing and publishing manufacturing sector using the above questions as topic headings.

Keep the brief report in your portfolio.

And now let's investigate the textiles, clothing and footwear industries.

➤ Textiles, clothing and footwear industries

◆ The textile industry

This industry manufactures a wide variety of goods from tablecloths to lace.

Let's look at a table of the top 20 UK textile companies in order of their turnover for the period ending 1992 or 1993.

No	UK Company	Turnover (£ millions)	Number of employees
1	Coats Viyella PLC	2 444	78 097
2	Courtaulds Textiles PLC	923	21 900
3	Baird [William] PLC	523	16 512
4	Dawson International PLC	432	11 973
5	Hartstone Group PLC (The)	370	4 465
6	Scapa Group PLC	347	6 422
7	Ashley [Laura] Holdings PLC	300	4 697
8	Readicut International PLC	235	3 859
9	Dewhirst Group PLC	182	6 621
10	Sherwood Group PLC	153	3 385
11	Levi Strauss (UK) Ltd	137	1 407
12	Pepe Group PLC	136	736
13	Lamont Holdings PLC	136	1 618
14	Allied Textile Companies PLC	127	2 306
15	Gent [SR] PLC	120	3 157
16	Claremont Garments (Holdings) PLC	114	3 824
17	Alexon Group PLC	108	3 970
18	Stirling Group PLC	101	3 376
19	Casket PLC	97	771
20	MCD (UK) Ltd	96	593

Top 20 UK textile companies
Source: *The Times 1000*, 1995 (Times Books)

Let's have a closer look at the individual textile industries.

The **wool textile industry** is one of the largest in the world, with the main manufacturing areas being West Yorkshire, Scotland, and the West of England.

The **cotton textile industry** involves manufacturing products by the processes of spinning and weaving.

The **linen industry** is centred in Northern Ireland and Scotland. Lightweight linen products such as household furnishings and textiles are manufactured in Northern Ireland, while heavyweight canvas products such as tents and tarpaulins are manufactured in Scotland.

The **man-made fibre industry** is dominated by just a few large companies. The main fibres produced are rayon, nylon and polyester.

The **carpet industry** is traditionally divided into the woven and non-woven sectors. Examples of woven carpets are Axminster and Wilton. The high quality of woven carpets makes the UK a leading manufacturer world-wide.

Products from the **jute industry** are used in the manufacture of carpets, ropes, packaging, and upholstery.

The **knitwear and hosiery industry** is made up of many small companies centred mainly in Scotland and the East Midlands. They produce underwear, stockings, tights, socks, and gloves.

The **cordage industry** is one of the largest in Europe and manufacturers nets, ropes, twine and netting.

Let's follow that with a look at the import and export trade of the UK textile industry:

Textile yarns and fabrics				
Year	Import trade (£ millions)	% of total UK import trade	Export trade (£ millions)	% of total UK export trade
1992	3 941	3.1	2 458	2.3

And what about the number of employees working in the UK textiles industry?

Textiles			
Period	No of employees	% of all manufacturing industry employees	% of all those in employment
Dec 1993	206 800	4.9	1.0

Now let's turn to the clothing industry.

◆ The clothing industry

The clothing industry is one of the largest in Europe, and it is dominated by thousands of small companies. It is very labour intensive, and mainly consists of cutting-out and making-up the garment using a sewing machine together with some finishing processes.

Let's have a look at the import and export trade of the UK clothing industry:

CLOTHING				
Year	Import trade (£ millions)	% of total UK import trade	Export trade (£ millions)	% of total UK export trade
1992	4 478	3.6	2 084	1.9

And how many employees work in the UK clothing, hats, gloves and fur goods industry?

COTHING, HATS, GLOVES AND FUR GOODS			
Period	No of employees	% of all manufacturing industry employees	% of all those in employment
Dec 1993	180 900	4.3	0.9

Right, let's take a look at the footwear industry.

◆ The footwear industry

This industry is similar to the clothing industry in that it is mainly made up of small companies where the footwear material is cut-out, made-up and then finished.

Now let's take a look at the import and export trade of the UK footwear industry:

FOOTWEAR				
Year	Import trade (£ millions)	% of total UK import trade	Export trade (£ millions)	% of total UK export trade
1992	1 152	0.9	341	0.3

And how many employees work in the footwear industry?

FOOTWEAR			
Period	No of employees	% of all manufacturing industry employees	% of all those in employment
Dec 1993	33 600	0.8	0.2

This task is carried out in conjunction with tasks 1, 2, 3, 4, 5, and 7

TASK 6

An investigation into the UK textiles, clothing and footwear manufacturing sector

To carry out this investigation, use all the resources available to you – such as the library, magazines, journals, teachers, tutors, local textiles, clothing and footwear companies, and 'yellow pages'.

- Into what subsectors is the UK textiles, clothing and footwear manufacturing sector divided?
- What is the typical range of products manufactured by the UK textiles, clothing and footwear manufacturing sector?
- Who are the main companies in the UK textiles, clothing and footwear sector?
- Is the UK textiles, clothing and footwear manufacturing sector dominated by just a few large companies or are there lots of small and medium-sized companies?
- Where are the main textiles, clothing and footwear companies located regionally in the UK?
- How many workers are employed in the UK textiles, clothing and footwear manufacturing sector?
- What are the import and export trade figures for the UK textiles, clothing and footwear manufacturing sector?

Produce a brief report on the UK textiles, clothing and footwear manufacturing sector using the above questions as topic headings.

Keep the brief report in your portfolio.

REASONS FOR THE LOCATION OF MANUFACTURING COMPANIES

- raw materials
- labour and skills supply
- energy supply
- local and national government incentives
- infrastructure

1.3 REASONS FOR THE LOCATION OF MANUFACTURING COMPANIES

There are a number of factors that affect where a factory is situated or built. Normally one factor outweighs the others when deciding where to locate the factory.

Let's begin by looking at raw materials.

➤ Raw materials

In many industries the access to raw materials is very important in deciding where to locate a manufacturing organisation. Beet-sugar refineries, for example, are located at the beet fields and paper-making companies at the conifer forests. The main reason for doing this is to reduce the transportation costs of the raw materials. Broadly speaking, heavy industries, such as iron and steel making, shipbuilding, cement and brick making, tend to locate themselves near to the raw materials which they use in such large quantities.

➤ Labour and skills supply

Different types of skilled labour can be found regionally across the UK. This is mainly because each region has many different traditional and heavy industries requiring labour supplies with traditional skills. Examples of this are the textiles and potteries industries, the automobile and aircraft industries, and the shipbuilding industry.

Workers tend to stay in the same region because of financial constraints and social considerations.

Modern industries are not necessarily attracted to regions of high unemployment because of the problems of finding the labour supply with the correct skills.

➤ Energy supplies

In order to reduce energy costs, many industries, particularly heavy industries like iron and steel-making, tend to be located close to their essential energy supplies, such as coal.

Industries which are dependent upon electricity, gas and oil can locate themselves practically anywhere in the UK.

➤ Incentives

In areas of high unemployment, local or national government and European Union (EU) incentives are offered to try to induce manufacturing organizations to locate in those areas. The incentives include financial assistance such as low interest loans, low rent levels, and special grants. This includes Enterprise zones and the European Development Fund.

➤ Infrastructure

◆ Transport and distributive costs

Transport costs are incurred whenever raw materials and component parts are brought into the factory and when the finished products are distributed to the wholesalers and customers. If the raw material is expensive to transport compared with the finished product, then the factory is more likely to be located near to the raw material. Conversely, if the finished product is heavy and/or bulky, as in brewing, then the factory is more likely to be located near to the market.

The level of transport costs depends upon the following factors:

- quantity;
- weight;
- bulkiness;
- fragility;
- perishability.

Factories should ideally be located close to first-class transport facilities including road, rail, sea, and air.

◆ Access to markets

Many manufacturing organisations are located in the south-east of England in order to be closer to the consumer markets of Europe. Eurofreight terminals have been set up to provide manufacturing companies with access to the European consumer markets.

EXAMPLES

Here are two examples of manufacturing companies who have located themselves in areas of the UK for particular reasons.

Coca Cola sited a soft drinks factory near Wakefield in Yorkshire for the following reasons:

- the quality of the water;
- the motorway network for distribution;
- the 'track record' of the local workforce.

Toyota sited a car factory near Derby for the following reasons:

- the 'track record' of the local workforce;
- the motorway and rail network for distribution.

This task is carried out in conjunction with tasks 1, 2, 3, 4, 5, and 6

TASK 7

An investigation into the reasons for the location of manufacturing companies

To carry out this investigation use all the resources available to you.

How do the following influence the decisions about where companies locate their operation?

- resources such as raw materials, labour and skills supply, and energy supply;
- local and national government incentives;
- infrastructure such as access to markets and transportation costs.

Write a brief report that considers, in general terms, how resources, incentives and infrastructure influence decisions about where companies locate their operations, and give two examples of companies which decided to locate in particular areas for different reasons.

Bring together the brief reports from tasks 1–7 inclusive to produce a report titled: 'Investigation of the importance of manufacturing to the UK economy'.

Keep the final report in your portfolio.

KEY PRODUCTION STAGES

- material preparation
- processing
- assembly
- finishing
- packaging

1.4 KEY PRODUCTION STAGES

The manufacturing of a product can be broken down into these five key production stages.

Let's start by discussing the first production stage – called material preparation.

➤ Material preparation

The preparation of materials before processing operations are carried out is extremely important because the quality of the manufactured product is directly related to the degree of preparation carried out.
   Examples of preparation are:

- washing foodstuffs in the food processing industry;
- cutting materials in the textile industry;
- cleaning the mould in the plastics industry;
- degreasing the metal in the engineering industry.

➤ Processing

Processing is the conversion of materials from one form into another.
   Examples of processing are:

- compression moulding in the plastics industry;
- baking of rolls in the food industry;
- machining of metal in the engineering industry.

➤ Assembly

Assembly is the joining together of the various elements that make up a product.
   Examples of assembly are:

- stitching of fabrics in the clothing industry;
- soldering of components in the electrical and electronics industry;
- glueing of parts in the furniture industry.

➤ Finishing

Once materials have been processed and assembled they quite often require some form of finishing.

Finishing is normally applied to:

- make the product attractive;
- give protection.

Examples of finishing are:

- varnished wood;
- glazed food;
- spray-painted metal;
- electroplated metal;
- shrink-resistant wool;
- anti-static finishes on man-made fibres;
- water-repellent materials;
- stain-resistant carpets;
- flame-resistant fabrics.

➤ Packaging

Packaging is used to:

- identify the product;
- carry it through the distribution system to the customer.

Its basic functions are to:

- maintain product quality;
- meet the customer need;
- communicate and sell;
- conform with legal and regulatory aspects;
- protect against mechanical forces;
- protect against contamination;
- protect against climatic conditions;
- contain the product.

Typical packaging materials are:

- paper and board;
- plastics: flexible or rigid;
- glass: lightweight but strong;
- cans: low carbon steel protected by a thin layer of tin.

On the opposite page we can see the production stages in producing an ice cream jam roll.

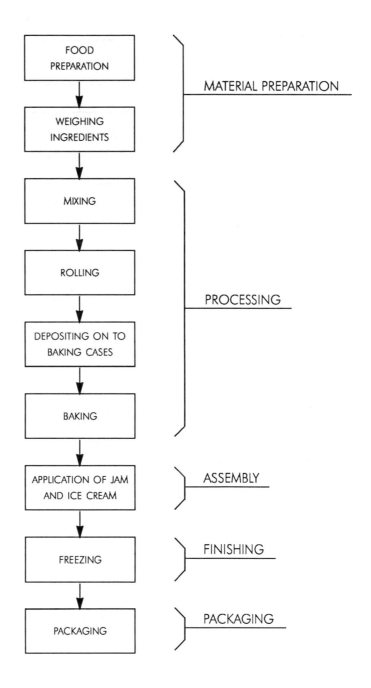

Flow diagram of the key production stages of ice cream jam roll production

On the next page you will find another simple example – this time illustrating the production stages in the manufacture of a metal tray.

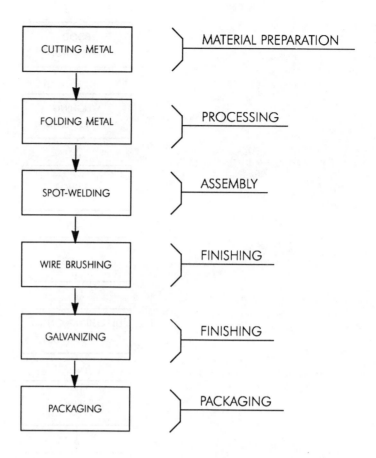

Flow diagram of the key production stages of a metal tray

This task is carried out in conjunction with tasks 9, 10, 11, and 12

KEY STAGES OF PRODUCTION

- material preparation
- processing
- assembly
- finishing
- packaging

TASK 8

An investigation into the key stages of production in the manufacture of a product

To carry out this investigation, use all the resources available to you – such as the library, magazines, journals, teachers, tutors, local companies, and 'yellow pages'.

Select one out of the following 12 products:

- stool
- skirt
- biscuit
- paint
- magazine
- compact disc
- pencil
- coat hanger
- cheese
- can of cola
- light bulb
- hand-made shoes

Find out using the resources available to you how your selected product is manufactured, and from this describe what occurs, in general terms, at each of the key stages of production.

Produce a brief report, including a flow diagram, of the key stages of production required in the manufacture of your selected product.

Keep the brief report in your portfolio.

SCALES OF PRODUCTION

- continuous flow or line
- repetitive batch
- small batch or job

1.5   SCALES OF PRODUCTION

The production output volumes of manufacturing organizations vary from one to many thousands in a given time, depending upon the product being manufactured: from a ship to a light bulb, or from a piece of furniture to a bottle of milk.

The production systems used to manufacture such a range of products are categorized into three broad areas: continuous flow or line, repetitive batch and small batch or job.

➤ Continuous flow or line production system

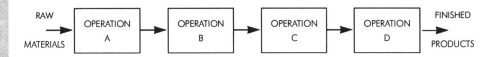

RAW MATERIALS → OPERATION A → OPERATION B → OPERATION C → OPERATION D → FINISHED PRODUCTS

DEFINITION
Continuous flow or line production system: one in which products are manufactured within a continuous process, moving from one operation to the next in a fixed pre-determined sequence.

Typical product outputs

- motor cars
- petrochemicals
- paper-making
- light bulbs
- domestic gas
- sugar refining
- bottles of milk

Characteristics

- long production runs on the same type of product;
- rigid product specifications;
- system is usually inflexible because it is unable to accept product changes very easily;
- high investment costs;
- highly specialized plant;
- production process equipment is arranged in operation sequence;
- stoppages in the flow of production have to be rectified immediately because the production line could be immobilized;
- materials are normally transferred from one operation to the next via mechanized systems or pipes.

➤ Repetitive batch production system

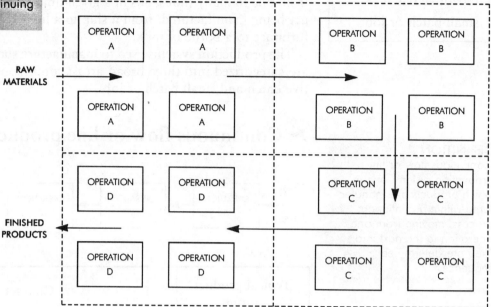

Typical product outputs	Characteristics
• electric motors • switchgear • furniture • kitchen and bathroom domestic products • screws • castings • vacuum cleaners • television sets • car components • glassware • clothes • office items • books • bakery products • bottles of beer • confectionery • washing machines • radios	• similar types of production process equipment are placed in groups or batteries; • general purpose machine tools and equipment are used to undertake the manufacture of a wide range of products; • expensive tooling arrangements to provide for changes in set-up; • intermittent movement of products from one group of machines to the next; • normally long production period because of the waiting time between operations; • extremely flexible being able to accommodate many different types of product; • batch sizes vary widely; • variations in production and consumption rates lead to the surplus products being stored; • most common method in UK manufacturing.

➤ Small batch or job production system

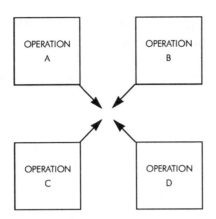

<table>
<tr><td>OPERATION A</td><td>OPERATION B</td></tr>
<tr><td>OPERATION C</td><td>OPERATION D</td></tr>
</table>

Typical product outputs

- ships
- hand-made shoes
- prototype equipment
- oil rigs
- purpose-built piece of equipment
- hand-made clothing
- wedding cake
- wedding dress
- specially modified equipment
- hand-made furniture

Characteristics

- wide range of general purpose machines and equipment;
- highly skilled and versatile workforce;
- jobs are tendered for;
- facilities are centred around a static product;
- fluctuating demand for specialized labour;
- good design organization required;
- high level of skills required to interpret the design and specification of the job
- accurate storekeeping required;
- sometimes a number of site engineers are required.

Prototype equipment
Used in the pre-production stage to determine whether a design functions to specification.

This task is carried out in conjunction with tasks 8, 10, 11, and 12

SCALES OF PRODUCTION

- continuous flow or line
- repetitive batch
- small batch or job

TASK 9

An investigation into the scales of production in the manufacture of a product

To carry out this investigation, use all the resources available to you – such as the library, magazines, journals, teachers, tutors, local companies, and 'yellow pages'.

Continue with the product that you selected in task 8 and investigate the scale of production used to manufacture it.

Produce a brief report summarizing the scale of production used to manufacture your selected product.

Keep the brief report in your portfolio.

1.6 PRODUCTION SYSTEMS

The main function of a production system is to convert raw materials into a finished product that can be sold for profit.

Let's have a look at the typical inputs and outputs of a production system.

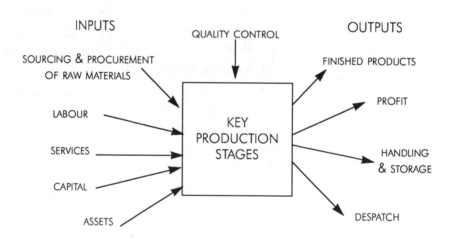

A production system

INPUTS
• sourcing and procurement of raw materials
• labour
• services
• capital
• assets

➤ Inputs

◆ Sourcing and procurement of raw materials

Sourcing is finding suitable sources for obtaining materials, and procurement is the acquiring of the materials from these sources.

Raw materials include the basic items such as cotton, flour and polymers, and also bought-in components manufactured by other companies such as tyres used in the car manufacturing industry.

◆ Labour

Includes employees who are involved both directly and indirectly in the production process, such as operatives and administrative staff respectively.

◆ Services

Services include electricity, gas, oil, water, and drainage. Mail delivery and collection, facsimile transmission, electronic mail and telephones are further examples of services.

◆ Capital

Capital comes from the sale of the finished product and can also be raised by borrowing money from shareholders and banks. Capital can be used to buy new machinery and equipment, expansion of the business, and for the purchase of raw materials and labour.

◆ Assets

Assets include the premises, machinery and equipment, tools, vehicles, spares and materials.

The conversion process of the materials into a finished product is performed during the key stages of production which we met in section 1.4.

➤ Key stages of production

The key stages of production are:

• material preparation
• processing
• assembly
• finishing
• packaging

Finally, let's look at the outputs of a production process.

OUTPUTS

- finished products
- profit
- handling
- storage
- despatching

➤ Outputs

◆ Finished products

The final output from the conversion process.

◆ Profit

The sum of money left over after all expenses have been paid out of the sales revenue. Expenses include paying for materials, labour and services, and for the conversion process and delivery of the finished product to the customer.

◆ Handling

The administration required to receive orders and prepare for despatch of the finished product.

◆ Storage

The finished product is stored in buildings such as warehouses until required by customers.

◆ Despatching

The sending off of the finished product to the warehouse or customer.

Quality control is the measurement, recording and maintenance of the standards of production and is discussed in section 1.7.

Let's produce a flow diagram of a typical production system.

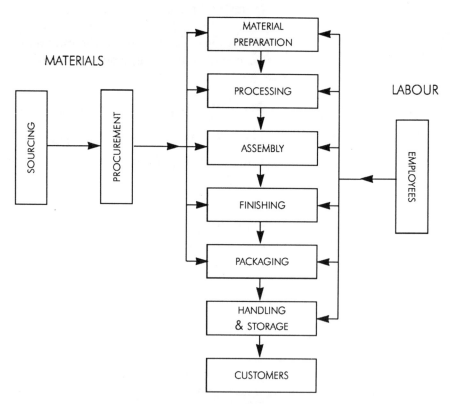

Flow diagram of a typical production system

Here is a flow diagram of the production system for the manufacture of the metal tray.

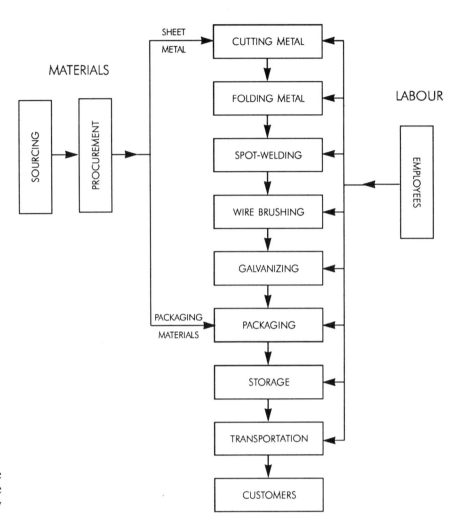

Flow diagram of the production system for the manufacture of a metal tray

This task is carried out in conjunction with tasks 8, 9, 11, and 12

PRODUCTION SYSTEM

• sourcing and procurement
• labour
• processing
• handling and storage

TASK 10

An investigation into production systems

To carry out this investigation, use all the resources available to you – such as the library, magazines, journals, teachers, tutors, local companies, and 'yellow pages'.

Continue with the product that you selected in task 8 and investigate the production system used to manufacture it.

Produce a flow diagram of the production system required to manufacture your selected product.

Keep the flow diagram in your portfolio.

1.7 QUALITY CONTROL

Let's begin by looking at quality assurance.

➤ Quality assurance

What is quality assurance?

It is the achievement of the required quality standards of the finished product before despatching it to the customer or storage. It is achieved by creating and maintaining a quality system within the manufacturing organization.

British Standard 5750 is the UK standard for quality management systems, and a manufacturing organization wishing to obtain the BS5750 kite mark has to have its quality systems assessed by an independent external organization. If it conforms to the national standards then it is awarded the BS5750 certificate and entered on a register held by the Department of Trade and Industry (DTI).

Now let's take a look at quality control.

➤ Quality control

What is the role of quality control?

The role of a quality control system is to measure, record and maintain the standards of production.

DEFINITION
Defective product: one which has a fault or flaw.

Quality control is achieved by establishing the expected quality standards required by the customer, and then planning to achieve those standards. As the product is manufactured it is inspected and if necessary corrective action is taken to ensure that the expected quality standards are met. Many manufacturing organizations are fully committed to manufacturing products 'RIGHT FIRST TIME', since defective products cost the company money.

What is RIGHT FIRST TIME?

This is directly concerned with getting the quality of a product right first time. Manufacturing a quality product is the result of quality manufacturing and quality control. Quality cannot be inspected into a product: it has to be correctly made from the start – that is 'right first time'.

The quality control department is organized to inspect the quality of the product.

Let's look at the points at which the product is inspected during its manufacture.

POINTS OF INSPECTION

- incoming raw materials and parts
- material preparation, processing and assembly
- finished product

DEFINITION
Acceptance sampling: inspecting a sample of the materials and parts received.

◆ Points of inspection

Let's begin by looking at the inspection of incoming raw materials and parts.

◆ Incoming raw materials and parts

If a manufacturing organization uses a particular material or parts not to the correct specification in the production process then the whole batch of products may have to be scrapped. To try and prevent this the company must ensure that only materials and parts which conform to the given specifications are accepted from the suppliers. Most manufacturing organizations carry out acceptance sampling on the incoming materials and parts, and have a vendor rating for the supplier which takes into account various quality related factors, such as the percentage of acceptable items received in the past and the percentage of warranty claims which have been traced to defective items supplied by the vendor.

The product is inspected at various stages of its manufacture.

◆ Material preparation, processing and assembly

If one of these production stages introduces a defect then the whole batch may possibly be scrapped. To ensure that only products which conform to the specifications, which will be discussed in Chapter 2, are manufactured, inspections are carried out often using control charts, which will be discussed in Chapter 3. The number and location of these inspections will depend upon the probability of a defect occurring and the cost of carrying out the inspections.

The final inspection of the product is very important.

◆ Finished product

The finished product is quality inspected to ensure that it conforms to the product specifications by using methods such as acceptance sampling, which will be discussed in Chapter 3. This quality inspection is very important because if a defective product is not identified at this stage then it will be passed on to the customer.

Quality control departments need to establish well-defined procedures for each point of inspection. These should include:

- selection and inspection of products;
- recording and analysis of data;
- reworking, scrapping or downgrading of defective products;
- feedback of information.

◆ Quality indicators

A quality indicator is a variable or an attribute of a product that can be measured or assessed respectively. The data obtained is then compared with the product's specification to give an indication of its quality.

Here are some examples of quality indicators:

- weight;
- volume;
- size;
- functionality;
- appearance;
- taste;
- sound;
- smell;
- touch.

Quality indicators will be discussed in more detail in Chapter 3.

Now let's look briefly at the inspection and testing methods that are carried out on the product once it has been selected for inspection.

◆ Inspection and testing methods

Inspection and testing methods are an important activity in ensuring that a quality product has been manufactured.

Here are some examples of inspection and testing methods:

- visual;
- mechanical;
- electronic and electrical;
- chemical analysis;
- expert scrutiny.

Inspection and testing methods will be discussed in detail in Chapter 3.

◆ Data recording formats

The results of the inspection and testing of the manufactured products are recorded in either written form or using a computer, and displayed in graphical or tabular form.

Data recording formats will be discussed in more detail in Chapter 4.

Now let's add the quality control points to the flow diagram of a typical production system.

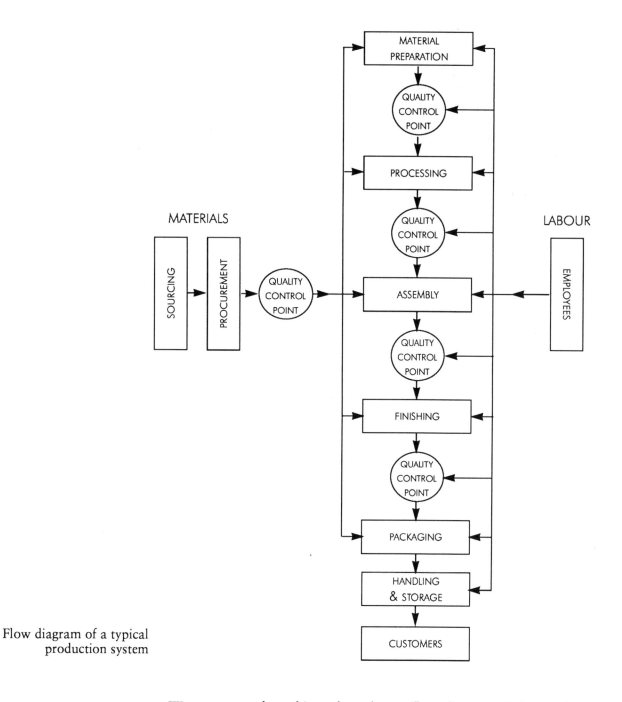

Flow diagram of a typical
production system

We can now adapt this and produce a flow diagram of the production
system for the manufacture of the metal tray. You will find this on the
next page.

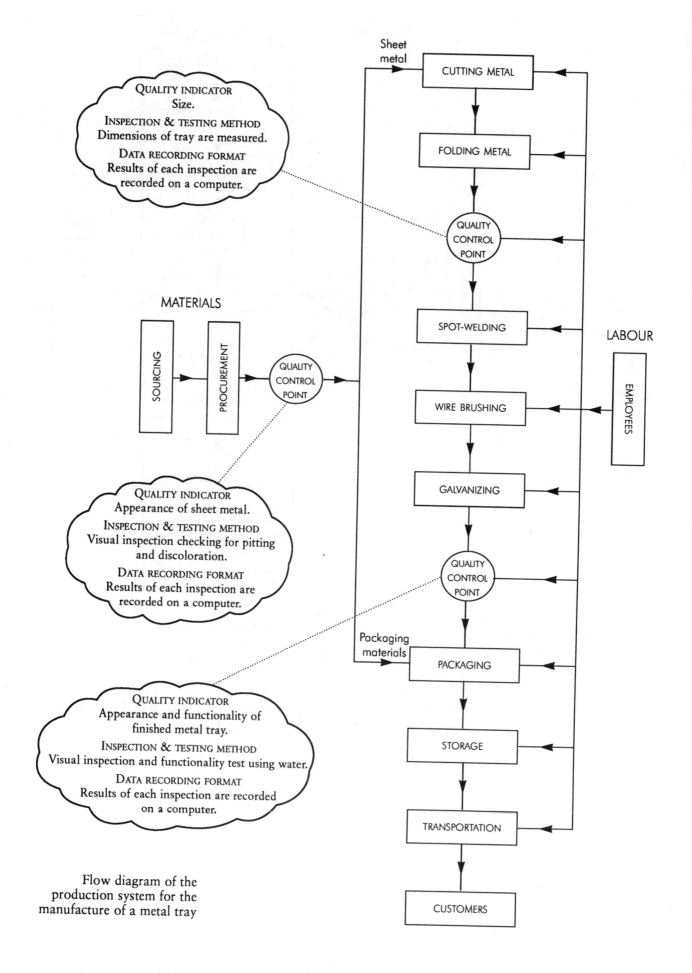

Flow diagram of the production system for the manufacture of a metal tray

And here is a flow diagram of the production system for the manufacture of a sports skirt.

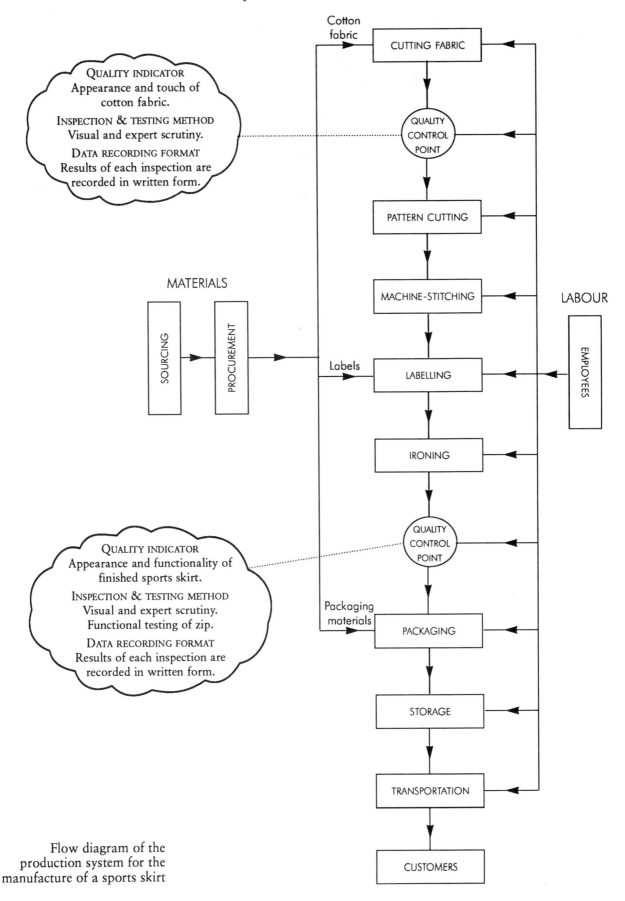

QUALITY INDICATOR
Appearance and touch of cotton fabric.

INSPECTION & TESTING METHOD
Visual and expert scrutiny.

DATA RECORDING FORMAT
Results of each inspection are recorded in written form.

QUALITY INDICATOR
Appearance and functionality of finished sports skirt.

INSPECTION & TESTING METHOD
Visual and expert scrutiny.
Functional testing of zip.

DATA RECORDING FORMAT
Results of each inspection are recorded in written form.

Cotton fabric

CUTTING FABRIC

QUALITY CONTROL POINT

PATTERN CUTTING

MACHINE-STITCHING

MATERIALS

SOURCING

PROCUREMENT

Labels

LABELLING

LABOUR

EMPLOYEES

IRONING

QUALITY CONTROL POINT

Packaging materials

PACKAGING

STORAGE

TRANSPORTATION

CUSTOMERS

Flow diagram of the production system for the manufacture of a sports skirt

This task is carried out in conjunction with tasks 8, 9, 10, and 12

TASK 11

An investigation into quality control

Use all the resources available to you in order to carry out the following investigation.

Continue with the product that you selected in task 8.

Using the flow diagram that you produced in task 10, decide where you think the quality control points should be positioned within the production system in order that the product is manufactured to its specification.

Add quality control points to your flow diagram and make general comments at each of these points including:

- Type of quality indicator to be used.
  For example: weight, volume, size, functionality, appearance, taste, sound, smell, touch.
- Inspection and testing method to be used.
  For example: visual, mechanical, electronic and electrical, chemical analysis, expert scrutiny.
- Data recording format to be used.
  For example: written form or using a computer.

Keep the flow diagram in your portfolio.

1.8   EFFECTS OF CHANGING THE SCALES OF PRODUCTION

These are summarized in the table on the page opposite.

PLANT LAYOUT	WORK PRACTICES	COST	NUMBER OF PRODUCTS MANUFACTURED
(A) CONTINUOUS FLOW OR LINE PRODUCTION SYSTEM RAW MATERIALS → OP A → OP B → OP C → OP D → FINISHED PRODUCT Products are manufactured within a continuous process, moving from one operation to the next in a fixed, predetermined sequence.	CHANGING TO REPETITIVE BATCH SYSTEM Supervision, training and instructions must be given to employees on the introduction of new methods and processes. Longer production periods because of the waiting time between operations.	CHANGING TO REPETITIVE BATCH SYSTEM General-purpose machine tools and equipment need to be installed.	CHANGING TO REPETITIVE BATCH SYSTEM Reduction in product output.
	CHANGING TO SMALL BATCH OR JOB SYSTEM Supervision, training and instructions must be given to employees on the introduction of new methods and processes. Highly skilled and versatile workforce. Long production periods.	CHANGING TO SMALL BATCH OR JOB SYSTEM General-purpose machine tools and equipment need to be installed.	CHANGING TO SMALL BATCH OR JOB SYSTEM Large reduction in product output.
(B) REPETITIVE BATCH PRODUCTION SYSTEM RAW MATERIALS / FINISHED PRODUCTS OP A OP A / OP B OP B OP A OP A / OP B OP B OP D OP D / OP C OP C OP D OP D / OP C OP C Similar types of production process equipment are placed in groups or batteries.	CHANGING TO CONTINUOUS FLOW OR LINE SYSTEM Supervision, training and instructions must be given to employees on the introduction of new methods and processes. Short production periods.	CHANGING TO CONTINUOUS FLOW OR LINE SYSTEM High investment costs because it is specialized plant.	CHANGING TO CONTINUOUS FLOW OR LINE SYSTEM Increase in product output.
	CHANGING TO SMALL BATCH OR JOB SYSTEM Supervision, training and instructions must be given to employees on the introduction of new methods and processes. Highly skilled and versatile workforce. Long production periods.	CHANGING TO SMALL BATCH OR JOB SYSTEM No machine investment costs.	CHANGING TO SMALL BATCH OR JOB SYSTEM Large reduction in product output.
(C) SMALL BATCH OR JOB PRODUCTION SYSTEM OP A → ← OP B ↓ ↓ OP C → ← OP D Product (or products) remain stationary, and the workers, materials and equipment come to the product.	CHANGING TO CONTINUOUS FLOW OR LINE SYSTEM Supervision, training and instructions must be given to employees on the introduction of new methods and processes. Short production periods.	CHANGING TO CONTINUOUS FLOW OR LINE SYSTEM High investment costs because it is specialized plant.	CHANGING TO CONTINUOUS FLOW OR LINE SYSTEM Large increase in product output.
	CHANGING TO REPETITIVE BATCH SYSTEM Supervision, training and instructions must be given to employees on the introduction of new methods and processes. Longer production periods because of the waiting time between operations.	CHANGING TO REPETITIVE BATCH SYSTEM More general-purpose machine tools and equipment need to be installed.	CHANGING TO REPETITIVE BATCH SYSTEM Increase in product output.

This task is carried out in conjunction with tasks 8, 9, 10, and 11

TASK 12

An investigation into how changing the scales of production affects the organization of a manufacturing system

Using all the resources available to you, carry out the following investigation.

Continue with the product that you selected in task 8.

Begin with a small batch or job production system to manufacture your product.
• Draw the plant layout to manufacture your product.

Change the scale of production to a repetitive batch production system.
• Draw the new plant layout to manufacture your product.

What changes in work practices will occur?

What costs are involved in changing scales of production to repetitive batch from a small batch production system?

Is there an increase or decrease in production output of your product?

Change the scale of production to a continuous flow or line production system from a repetitive batch production system.
• Draw the new plant layout to manufacture your product.

What changes in work practices will occur?

What costs are involved in changing scales of production to continuous flow or line from a repetitive batch production system?
Is there an increase or decrease in production output of your product?

Produce a table showing the plant layout, changes in working practices, costs involved and changes in production output when you change from small batch or job production system to a repetitive batch production system and then to a continuous flow or line production system when manufacturing your product.

Bring together tasks 8–12 inclusive to produce a report titled: 'Investigate production systems'.

Keep the final report in your portfolio.

1.9 DEPARTMENTAL FUNCTIONS IN A MANUFACTURING ORGANIZATION

Let's investigate the different departmental functions within a manufacturing organization – beginning with quality control.

➤ Quality control

The quality control function is the:

- quality inspection of the products being manufactured to ensure that there is little chance of producing a defective product;
- gathering, recording and analysis of statistical quality records.

➤ Customer relations and marketing

The customer relations and marketing function comprises:

- contracting and supplying;
- generating, stimulating and facilitating sales;
- providing customers with information;
- dealing with customer enquiries and complaints.

➤ Research and development

Research can be either pure or applied.
Pure research is carried out with the intention of furthering basic knowledge in a particular subject.
Applied research is carried out with the intention of investigating and solving an identified problem.

The research and development function is to:

- discover new ideas, information, techniques and systems;
- develop these new ideas, information, techniques and systems into practical applications.

➤ Design

The design function is to:

- translate the developments from the research and development department into products acceptable to the market.

➤ Sourcing and procurement

The sourcing and procurement function is to:

- identify suitable sources of materials supply;
- purchase the supplies.

➤ Production

The production function is to:

- manufacture a product to the agreed specifications and quantities within a scheduled time;
- maintain the machines and equipment.

Loading in the warehouse

➤ Storage

The storage function is to:

- store the finished product in a known location, under ideal conditions, until it is required to be retrieved.

➤ Handling and distribution

The handling and distribution function is to:

- administer the movement of goods from storage to consumer;
- transport the goods from storage to consumer.

➤ Financial control and planning

The financial control and planning function is to:

- prepare budgets for the main areas of the organization such as research and development, wages, salaries, sales and marketing, and production;
- prepare a master budget for the organization;
- carry out financial accounting;
- carry out investment appraisal.

This task is carried out in conjunction with tasks 14, 15, and 16

DEPARTMENTAL FUNCTIONS

- quality control
- customer relations and marketing
- research and development
- design
- sourcing and procurement
- production
- storage
- handling and distribution
- financial control and planning

TASK 13

An investigation into the departmental functions within a manufacturing organization

Using all the resources available to you, carry out the following investigation.

Identify the departmental functions in a local large manufacturing organization

Produce a brief report describing in general terms the departmental functions in a local large manufacturing organization.

Keep the brief report in your portfolio.

1.10 DEPARTMENTAL RESPONSIBILITIES DURING THE MANUFACTURE OF PRODUCTS

The flow diagram shows the responsibility of departments during manufacture.

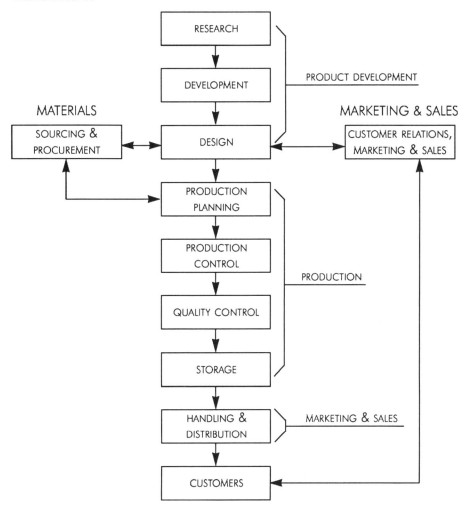

Flow diagram of departmental functions in the manufacture of a product

DEPARTMENT	RESPONSIBILITY DURING THE MANUFACTURE OF PRODUCTS
RESEARCH AND DEVELOPMENT/DESIGN	Setting a detailed specification for the product. Pre-production and prototype trials.
SOURCING AND PROCUREMENT	Identifying suitable suppliers for the materials and parts that are required for manufacturing the product. Negotiating to obtain the most competitive price. Ensuring that the material supplies are delivered to the right place on time.
MARKETING AND SALES	Pricing the product – dependent on quality and the market it is aimed at. Promoting the product – advertising; personal selling; publicity; sales promotion. Packaging the product. Administering the sales. Distributing the product.
PRODUCTION PLANNING	Scheduling the manufacture of the product, that is, planning the time to manufacture the product. Loading, that is, determining the work assigned to an operator or machine over a period of time, such as a day or a week.
PRODUCTION CONTROL	Despatching – preparation of all the necessary paperwork and documentation for manufacturing the product; – movement of materials, parts and finished goods in and out of stores. Progress – making the system run to schedule. Materials control – ensuring that the correct materials and parts are available at the right place on time.
QUALITY CONTROL	Inspecting the quality of the product at a number of pre-determined points between the materials and parts arriving at the factory and the finished product leaving the factory.
HANDLING AND DISTRIBUTION	Administering the movement of the product to the customer. Distribution of the product by direct sales, distribution chain or retail mail order.

This task is carried out in conjunction with tasks 13, 15, and 16

RESPONSIBILITIES
IN TERMS OF:

- product development
- production
- procurement
- marketing and sales

TASK 14

An investigation into the departmental responsibilities during the manufacture of products

Using all the resources available to you, carry out the following investigation.

Continuing with the manufacturing organization used in task 13, identify the departmental responsibilities during the manufacture of a product.

Produce a brief report describing in general terms the departmental responsibilities during the manufacture of a product in a local large manufacturing organization.

Keep the brief report and flow diagram in your portfolio.

1.11 ORGANIZATION CHARTS

The organization chart shows how an organization is structured. Everyone in an organization needs to know what each person's responsibility and function is. Organization charts quickly become out of date and should be regularly revised.

There are various types of organization charts: wheel charts, scalar chain charts, and the common 'family tree' charts.

First, let's have a look at a basic 'family tree' organization chart.
The chart on the next page will help you understand organizational relationships.

Work roles on the same horizontal level, such as Manager A and Manager B, tend to have the same degree of authority and to have closely related work.

A work role on a particular level has more authority than a work role vertically below it, such as Manager A and Supervisors A, B, and C.

Following the first basic organization chart you will find three other charts: they represent a small, medium and large manufacturing company respectively.

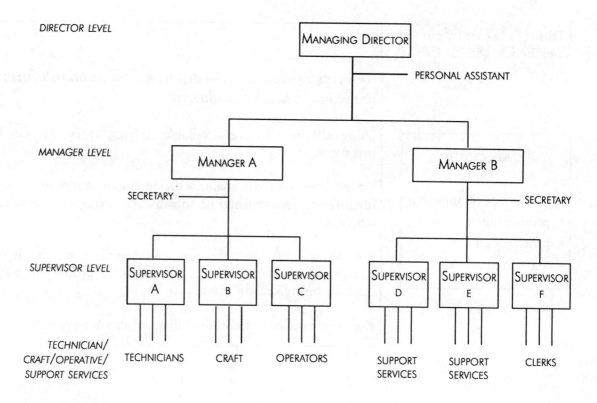

Basic organization chart

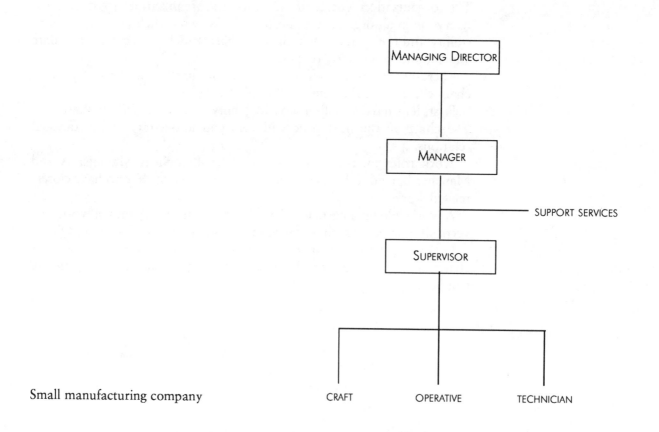

Small manufacturing company

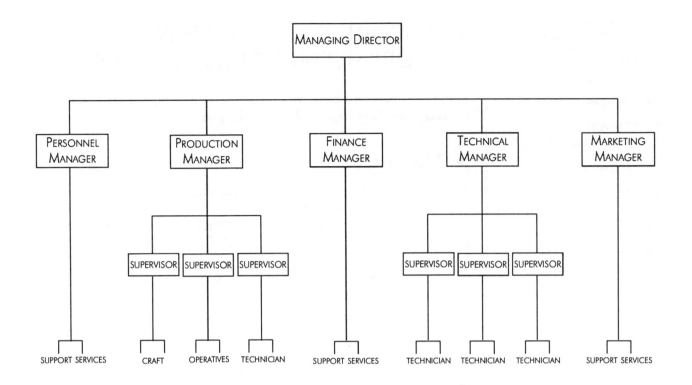

Medium-sized manufacturing company

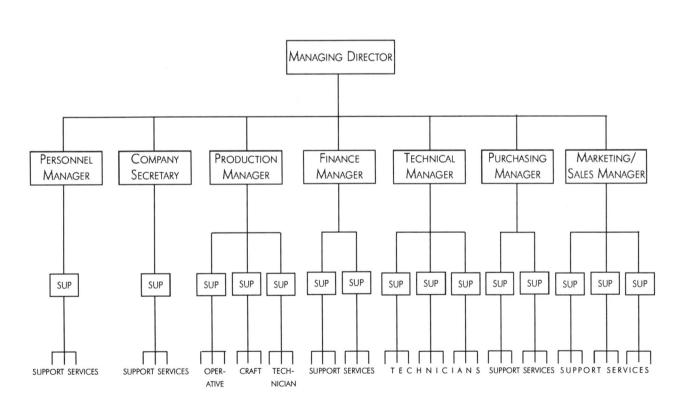

Large manufacturing company

SUP = SUPERVISOR

1.12 WORK ROLES WITHIN DEPARTMENTAL FUNCTIONS

There are many different work roles within a typical manufacturing organization. Let's look at them one by one.

➤ Managers

Usually a professionally qualified person who generally manages a division within a manufacturing organization.

Normally each manager will have a secretary, and, depending on the manager's function, will have a number of supervisors reporting directly to him or her.

Each manager is responsible to the managing director for the efficient running of their division.

➤ Supervisors

Normally an experienced person who directs or oversees the performance of a task or operation carried out by an operative or operatives within a section.

A supervisor is a junior manager who is accountable to the manager, and delegates tasks to the operatives within the section.

➤ Operators

Normally a trained person who operates a machine, instrument or tool, that is performs an operation.

Operators can be skilled, semi-skilled or unskilled, and are accountable to the supervisor for the standard of work that they turn out.

➤ Craftspeople

A person who has a skilled trade, or craft.

Normally a craftsperson has undergone an apprenticeship in that trade.

In some manufacturing sectors craftspeople have traditionally made products by hand.

➤ Technicians

Technicians are people skilled in a particular technical field.

Technicians are employed in such areas as the drawing office and quality control.

➤ Secretaries

A secretary is a person who handles correspondence, keeps records and carries out general office work for an individual or a group of individuals.

➤ Designers

A designer is a professional person who devises and executes designs, and would generally be employed in the technical division of a manufacturing organization.

➤ Maintenance engineers

A skilled worker who is trained in the maintenance of plant equipment and machines, and would generally be employed in the production division.

➤ Researchers

A researcher is a person who carries out a systematic investigation to establish facts or collect information on a subject. This involves research in all these areas: market, consumer, product, distribution and promotional.

A researcher is employed in the marketing and sales division.

➤ Marketing personnel

Personnel in the marketing division of a manufacturing organization who are involved in market research, advertising and sales promotion.

➤ Research and development staff

Graduate scientists and engineers who are generally conducting research in order to make new discoveries, and then develop them into commercial applications.

R & D staff are normally employed in the production division.

➤ Office staff

Clerical and administrative staff, typists and receptionists who are employed centrally in many manufacturing organizations.

➤ Drivers

A driver is a person who transports the manufactured products from the factory to the warehouses and customers. Normally a driver has a Heavy Goods Vehicle (HGV) licence in order to be able to drive large lorries such as articulated lorries. Drivers are employed in the distribution division.

Having discussed the personnel, now let's use charts to examine the organization of these work roles within departmental functions, starting with the quality control function.

➤ Quality control function

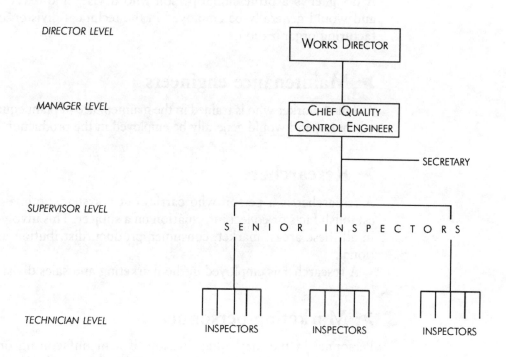

DIRECTOR LEVEL

WORKS DIRECTOR

MANAGER LEVEL

CHIEF QUALITY
CONTROL ENGINEER

SECRETARY

SUPERVISOR LEVEL

SENIOR INSPECTORS

TECHNICIAN LEVEL

INSPECTORS   INSPECTORS   INSPECTORS

➤ Customer relations and marketing function

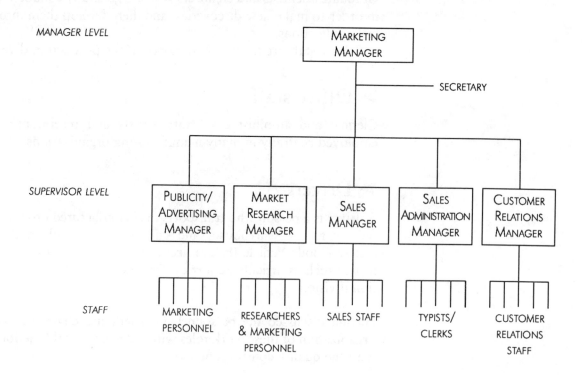

MANAGER LEVEL

MARKETING
MANAGER

SECRETARY

SUPERVISOR LEVEL

PUBLICITY/
ADVERTISING
MANAGER

MARKET
RESEARCH
MANAGER

SALES
MANAGER

SALES
ADMINISTRATION
MANAGER

CUSTOMER
RELATIONS
MANAGER

STAFF

MARKETING
PERSONNEL

RESEARCHERS
& MARKETING
PERSONNEL

SALES STAFF

TYPISTS/
CLERKS

CUSTOMER
RELATIONS
STAFF

➤ Research and development function

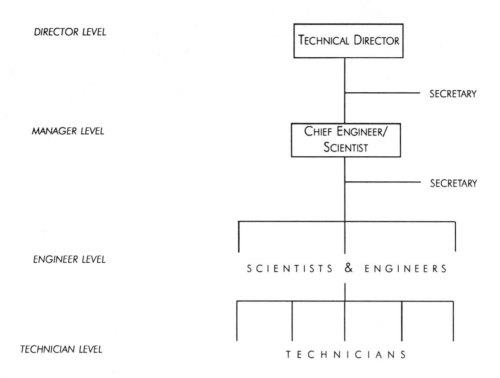

DIRECTOR LEVEL — TECHNICAL DIRECTOR — SECRETARY

MANAGER LEVEL — CHIEF ENGINEER/SCIENTIST — SECRETARY

ENGINEER LEVEL — SCIENTISTS & ENGINEERS

TECHNICIAN LEVEL — TECHNICIANS

➤ Design function

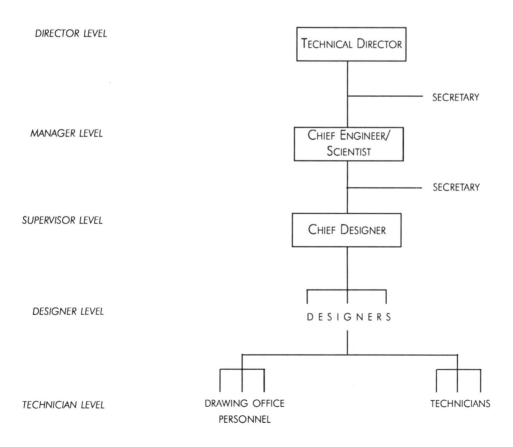

DIRECTOR LEVEL — TECHNICAL DIRECTOR — SECRETARY

MANAGER LEVEL — CHIEF ENGINEER/SCIENTIST — SECRETARY

SUPERVISOR LEVEL — CHIEF DESIGNER

DESIGNER LEVEL — DESIGNERS

TECHNICIAN LEVEL — DRAWING OFFICE PERSONNEL — TECHNICIANS

➤ Sourcing and procurement function

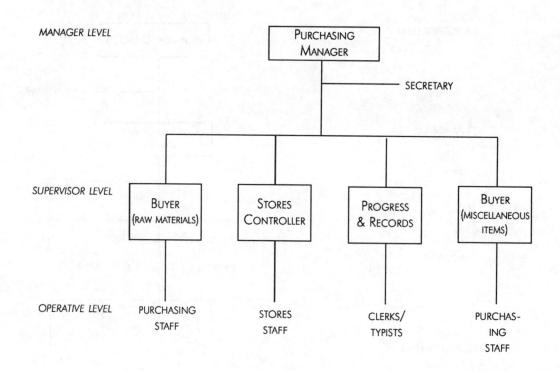

➤ Production function

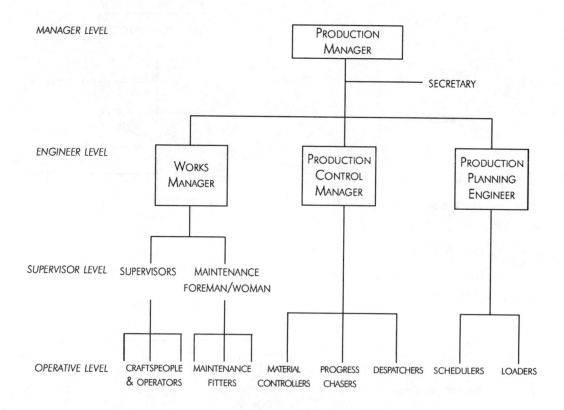

➤ Storage, handling and distribution function

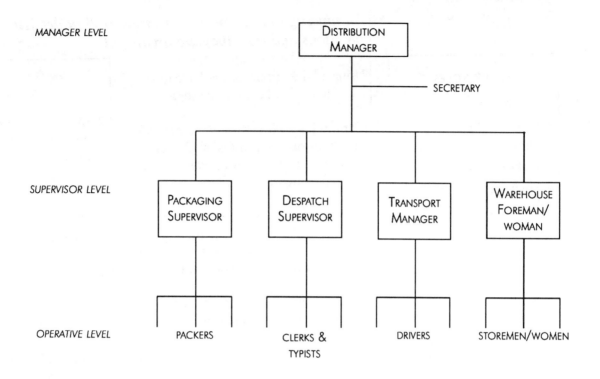

➤ Financial control and planning function

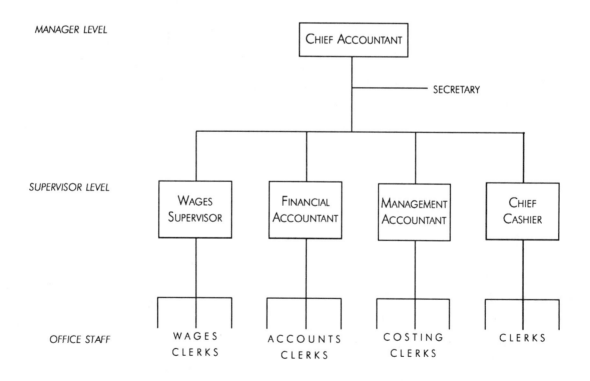

This task is carried out in conjunction with tasks 13, 14, and 16

TASK 15

An investigation into the work roles within departmental functions and how they are organized

Using all the resources available to you, carry out the following investigation.

Continuing with the manufacturing organization used in tasks 13 and 14, identify the main work roles within the departmental functions.

Produce a brief report describing in general terms the main work roles within the departmental functions of a local large manufacturing organization.

Produce an organization chart for each departmental function.

Keep the brief report and organization chart in your portfolio.

WORK ROLES
• managers
• supervisors
• shop-floor operators
• technicians
• engineers
• research and development
• office staff

DEPARTMENTAL FUNCTIONS
• quality control
• customer relations and marketing
• research and development
• design
• sourcing and procurement
• production
• storage
• handling and distribution
• financial control and planning

1.13 SKILL AND TRAINING REQUIREMENTS FOR WORK ROLES

Let's begin by looking at the different types of skill and training programme that are widely available.

SKILL AND TRAINING REQUIREMENTS
• vocationally specific
• general
• further
• higher

WORK ROLES
• managers
• supervisors
• shop-floor operators
• technicians
• engineers
• research and development staff
• office staff

Skill and training programmes

• NVQ/SVQ	vocationally specific/further education
• GNVQ	general/further education
• diplomas	general/further and higher education
• degrees	general/higher education
• company training certificates	vocationally specific
• GCSEs	general
• GCE A-levels	general
• modern apprenticeships	vocationally specific

➤ NVQ/SVQs

National Vocational Qualifications (NVQ) and Scottish Vocational Qualifications (SVQ) are about work. They are based on standards set by industry.

NVQ/SVQs are awarded by established bodies including City & Guilds, BTEC and RSA.

Each NVQ/SVQ is made up of a number of units which set out the standards that must be reached. A unit is like a mini-qualification. People can choose which units they do and are awarded certificates to show unit credits they have achieved.

NVQ/SVQs are placed in the NVQ framework according to the area of work and level they cover.

There are five levels in the framework, with level 1 the most basic.

Training can take place on the shop floor, in training centres or at a local college of further education.

Assessment can take place at work, and employees can get immediate credit for the competence they already have.

NVQ/SVQs provide a path for employees to gain promotion or employment in their area of work.

➤ GNVQs

General National Vocational Qualifications (GNVQs) are part of the framework of National Vocational Qualifications backed by the government, and are related to a particular area of work such as manufacturing.

Students taking a GNVQ in manufacturing are prepared for a career in manufacturing, or for higher education.

At present there are three levels available to pupils and students: Foundation level, Intermediate level, and Advanced level. In the next few years two other levels may be introduced which would run alongside, and in some cases replace, higher education courses such as HNDs and degrees.

GNVQs are awarded by City & Guilds, BTEC and RSA.

GNVQs are made up of units and candidates get credit for each unit as it is achieved, building up towards a full qualification.

GNVQs provide a path for entry into manufacturing at technician level, or for entry into higher education.

➤ Diplomas

Until recently there were three main types of diploma course for technicians: First Year Diploma, Ordinary National Diploma (OND) and Higher National Diploma (HND). They were all full-time courses with progression between them, designed to prepare students for technician positions in areas of work such as electrical or mechanical engineering.

Employed technician students attended part-time equivalent courses, the First Year Certificate, Ordinary National Certificate (ONC) and Higher National Certificate (HNC).

GNVQ (Intermediate) has now virtually replaced First Year Diploma.

GNVQ (Advanced) has replaced the Ordinary National Diploma (OND) in some colleges.

➤ Degrees

A relevant degree, with no experience, may provide an entry into junior management.

A relevant degree, with experience, may provide an entry into senior management.

➤ Company training certificates

These are attained after attending a course designed specifically for or by a company. Normally they are skills-oriented with minimum academic input, and hence specialized.

Depending on the depth of training involved, the duration of a company training course can vary from one day up to a month, or in some cases one day a week for a year.

The course is run at the place of work, training centre or college.

On completion of the training course, the student (employee) is awarded a company training certificate.

These types of training course could provide evidence towards achieving an NVQ.

➤ GCSEs

GCSEs are General Certificates of Secondary Education.

School-leavers achieving GCSEs at grade C or above in English, mathematics, science and one other subject may get an apprenticeship in a manufacturing organization, perhaps at a technical level.

School-leavers achieving GCSEs at grades D and E may get an apprenticeship in a manufacturing organization, perhaps at craft level.

➤ GCE A-levels

GCE A-levels are General Certificates of Education at Advanced level.

GCE A-levels passed in the appropriate subjects, at grade C and above, provide school-leavers with the opportunity of entering higher education to take their chosen degree or perhaps to apply for a junior management position in a manufacturing organization.

➤ Modern apprenticeships

Modern apprenticeships are a new approach to the development of young people leaving school. They are available in a wide cross-section of industries and open to both males and females between the ages of 16 and 17. Their purpose is to meet industry's widespread need for young people with high-quality skills.

Modern apprenticeships lead to a person achieving at least an NVQ level 3, with the possibility of going on to higher education. They are not based on time-serving and apprentices should be able to progress at the pace best fitted to their abilities and circumstances.

Let's now begin to look at the skill and training requirements for the main work roles within departmental functions.

➤ Junior manager

A professionally trained person who has completed a degree in a discipline such as production engineering, accountancy or design, and normally passed the relevant professional body's qualifying examinations, such as Engineering Council examinations. Most managers require training in interpersonal skills, that is handling people. They also need to be developing continuously to remain competent in their job because of the changes in technology and company structures that arise. Examples of junior managers are assistant works managers and transport managers.

➤ Senior manager

A senior manager's position in an organization would require a manager with relevant professional status, such as chartered engineer or chartered accountant, with the proven relevant experience and skills that come through many years of working in the world of manufacturing. Examples of senior managers are marketing managers and the chief accountant.

➤ Supervisors

A supervisor is a well-trained person who is normally qualified to Higher National Diploma level in an appropriate topic. It is common practice to have a supervisory qualification such as a Certificate in Supervisory Management.

Supervisory jobs are not all alike, even within the same company, and therefore a common training programme is not likely to achieve the required level of training and so it would need to be supplemented with other forms of development and training activities. A clearly defined job specification is a good starting point for successful training. Examples of supervisors are maintenance foremen/women and accounts supervisors.

➤ Shop-floor operators

An operative is a semi-skilled worker who has normally left school qualified up to General Certificate of Secondary Education (GCSE) level. Job training is provided on the shop floor. In addition, the operative normally attends the local Further Education (FE) college one day a week for skills improvement on an appropriate National Vocational Qualification (NVQ) course.

Modern production techniques often require workers to have a degree of flexibility which will enable them to move from one type of process operation to another as the need arises.

Examples of operatives are sewing machine operators and lathe operators.

➤ Technicians

A highly-trained person who has widely used skills, and has normally completed a BTEC Higher National Certificate (HNC) or GCE A-levels.

Technicians usually spend around three or four years being trained on the job, at the local FE college on day release, and on general training courses. Some technicians undergo an apprenticeship with their company. Most technicians are regarded as generalists rather than specialists.

Examples of technicians are design technicians, draughtspersons, and laboratory technicians.

➤ Engineers

Many engineers are university graduates, and if they have gained an honours degree on a course accredited by the relevant engineering institution, then they are on the first step towards chartered engineer status. After graduation, most engineers spend a period of at least two years being trained by the company; this involves a wide range of experiences, such as working in different departments or on different types of project. On completion of this initial training, the engineer might become responsible for a small project or perhaps manage a section.

After a minimum of seven years, including the time spent on the degree course, an engineer should have gained the professional experience to become a chartered engineer, and now be ready for senior management work roles.

Examples of engineers are design engineers, production engineers, and process (chemical) engineers.

➤ Research and development

The research and development department is made up of graduate engineers, physicists, mathematicians, chemists and biologists who normally have a minimum qualification of a 2.1 honours degree. They are innovative, ingenious and creative people who are normally involved in investigative work. They devise experiments to test their theories, take measurements to analyse the results, and finally compare them with the expected results. Development engineers, for example, take the researched product and decide how it will be manufactured.

Research and development engineers and scientists are provided with opportunities to develop both technical and managerial skills. During the first two years' training there is on-the-job training and instruction, project-based training, in-company courses, external courses, and induction courses together with appraisal and counselling.

Examples of research and development work roles are R & D engineers, and design and development engineers.

➤ Office staff

Office staff include work roles such as secretaries, clerical and administrative staff, telephonists and receptionists.

Secretarial staff require skills such as word processing and filing. These types of skill are learnt at secretarial colleges and colleges of FE on full-time courses, normally straight from school. During employment, further training skills, such as using new software packages including word processing, spreadsheets and databases, are taught on company courses and at the local college of FE.

Clerical and administrative staff require training in general office skills. These skills are normally specific to each organization and hence the majority of their training is carried out on the job.

Examples of office staff work roles are invoice clerks and typists.

This task is carried out in conjunction with tasks 13, 14, and 15

SKILL AND TRAINING
PROGRAMMES

- NVQ/SVQ
- GNVQ
- diplomas
- degrees
- company training
- certificates
- GCSEs
- GCE A-levels
- modern apprenticeships

WORK ROLES

- managers
- supervisors
- shop-floor operators
- technicians
- engineers
- research and development
- office staff

TASK 16

An investigation into the skill and training requirements for work roles

Using all the resources available to you, carry out the following investigation.

Continuing with the manufacturing organization used in tasks 13, 14, and 15, identify the main skill and training requirements for work roles.

Produce a brief report describing in general terms the main skill and training requirements for work roles within the departmental functions of a local large manufacturing organization.

Bring together tasks 13–16 inclusive to produce a report titled: 'Describe manufacturing organizations'.

Keep the final report in your portfolio.

1.14 LOCAL ENVIRONMENTAL IMPACT OF PRODUCTION PROCESSES

ENVIRONMENTAL
- human
- built
- natural

IMPACT
- quality of life
- local economy
- renewability of resources
- local geography
- pollution

The production processes used to manufacture products in the UK have a wide and varying impact on the local environment. These range from providing employment to polluting rivers, and from polluting the atmosphere to providing economic wealth.

Let's start by considering the effects of production processes on the quality of life.

➤ Quality of life

◆ Human

The manufactured goods produced by production processes such as cars, materials, computers, television sets, food and clothes have all generally improved the quality of life of most people in the UK.

The noise and odour from certain types of production process can seriously affect the quality of life of residents who live close to them.

◆ Built

The products produced for the construction industry provide houses, buildings and structures for local residents and the community. Roads and rail allow people to travel throughout the UK relatively quickly and easily.

◆ Natural

The construction of buildings, structures and roads in particular normally have an adverse effect on the local natural environment. Areas of natural beauty and havens of wildlife are destroyed forever by these constructions.

➤ Local economy

◆ Human

Manufacturers provide economic wealth and employment for the local community. This not only includes those who are directly employed by the manufacturing organization, but also those who provide supplies and services to the organization.

➤ Renewability of resources

◆ Natural

Many resources used in production processes are non-renewable. However, those that can be renewed must be able to be re-planted or bred.

For example, resources that are grown can be renewed by planting, such as wood, wheat, barley, corn, tea, rice, and rubber. Resources from the animal world can be renewed by breeding, for example hide, leather, meat, milk, and eggs.

➤ Local geography

◆ Natural

Local geography in both small and large communities is continually undergoing changes because of new developments occurring in their areas.

For example, with a steady decline in many of the heavy industries such as iron and steel production, and coal mining, there has been a policy in many communities around the UK to develop light industrial areas (estates) to provide local economic wealth and employment. These have changed the local geography wherever they have been established.

➤ Pollution

Pollutants are normally carried by air or water and can have an adverse effect on the human, built and natural environments. Let's begin by looking at how these environments are affected by air pollutants.

◆ Air pollution

Pollutants enter the atmosphere either directly from the production process or they are produced by chemical reactions in the atmosphere.

A major source of air pollution is the combustion of fuels such as oil and coal in production processes which give rise to emissions of smoke, and gases such as sulphur dioxide, nitrogen oxides, carbon dioxide and carbon monoxide, together with particles. **Sulphur dioxide** is a major contributor to acid rain which we will meet later in this chapter. **Nitrogen oxides** in high concentrations can reduce plant growth and damage sensitive crops, as well as contributing to the deposition of acid rain and the formation of ground level ozone. **Carbon monoxide** is toxic in high concentrations and can affect physical co-ordination, vision and judgement; it also contributes indirectly to global warming, as we will see later in the chapter.

> **Ground level ozone**
>
> Ozone, a colourless gas with a chlorine-like odour, naturally occurs in small quantities from ground level up to about 16 km. Increases in ozone concentrations can occur by chemical reactions of nitrogen oxides with other pollutants such as volatile organic compounds (VOCs). Increases in ground level concentrations of ozone can affect people's lungs and eyes, can reduce crop yields, and cause damage to natural vegetation.

Concentrations of lead and other elements such as arsenic, zinc, and iron can get into the atmosphere from the burning of coal and from metal works. These concentrations, particularly lead, collect in the environment and can affect people's health.

DEFINITION
Pollutant: a substance which is present at concentrations which cause harm or exceed an environmental quality standard.

DEFINITION
Acid rain: produced by various processes when acidic gases and particles are deposited on land and water.

DEFINITION
Toxic: poisonous.

DEFINITION
Global warming: an increase in the average temperature of the earth which is thought to be caused by the build-up of greenhouse gases.

Volatile organic compounds (VOCs) are chemical compounds which are emitted from a wide range of production processes, particularly those using adhesives and cleaning agents. VOCs readily evaporate but they can also contribute to air pollution by taking part in chemical reactions with nitrogen oxides forming ground level ozone.

◆ Water pollution

Water pollution generally results from accidental spillages of pollutants. Around 12% of all water pollution is caused by industry – as a direct result of industrial spillages and indirectly by the leaching (seeping) of contaminants from the storage of hazardous chemicals.

Water pollution is potentially harmful to both the freshwater and salt-water ecosystems, and to the people who come into contact with these systems.

Other forms of pollution are: noise, odour, and visual.

◆ Noise pollution

This is a form of pollution that can irritate and annoy anyone who comes into contact with it. People exposed to high noise levels for a prolonged period of time may have their health affected. Employees working in noisy environments, for example on production lines, may suffer some permanent loss of hearing. People living close to a noisy production process may find their quality of life affected.

Noise pollution from production processes can be minimized by planning the site layout and building design, constructing noise barriers, and limiting the number of hours the process is in operation.

◆ Odour pollution

Smell is a form of pollution which can annoy anyone who comes into contact with it. Odours and gaseous emissions are particularly pungent from processes such as chemical plants, and food processing establishments such as mushroom factories.

◆ Visual pollution

Many production processes, particularly in the heavy industry sector, do not visually enhance the local environment.

This task is carried out in conjunction with tasks 18, 19, and 20

TASK 17

An investigation into the local environmental impact of production processes

To carry out this investigation, use all the resources available to you – such as the library, magazines, journals, teachers, tutors, local companies, and 'yellow pages'.

Select two contrasting products with different scales of production. Here are some suggestions.

Continuous flow production system	Repetitive batch production system	Small batch/single item production system
paper	books	hand-made shoes
sugar	screws	wedding cake
bottles of milk	paint	customized car
light bulbs	cola drink	hand-made furniture
petrochemicals	bakery products	tailored clothing

Describe the local environmental impact of the production processes used to manufacture your two selected products.

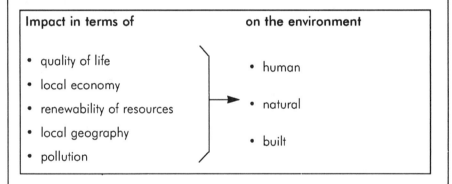

Impact in terms of	on the environment
• quality of life	
• local economy	• human
• renewability of resources	• natural
• local geography	• built
• pollution	

Produce a brief report on the local environmental impact of the production processes used to manufacture your two selected products.

Keep the brief report in your portfolio.

1.15 WASTE AND BY-PRODUCTS

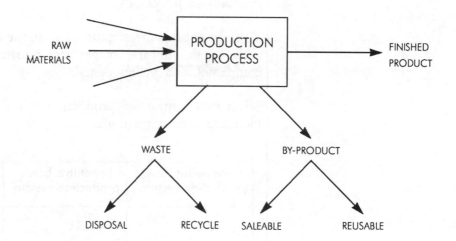

Difference between waste and by-product

➤ What is waste?

Waste is described as something that has no further use, and needs disposal. Waste from production processes, known as industrial waste, is either disposed of by waste disposal methods such as landfill or it is recycled.

Whilst some of these industrial wastes are relatively harmless, that is non-toxic, others are highly toxic. Major contributors to industrial waste are chemical plants, metal manufacturers and food manufacturers. Industrial waste is classified as controlled waste and is subject to waste disposal licensing under the Control of Pollution Act 1974. If the controlled waste contains substances which are a danger to life, then it is classified as **special waste** and subject to additional regulations designed to track its movement from production to disposal. Special wastes include poisons, corrosives, flammables, and prescription-only medicines.

A term you will come across is **hazardous waste**, which includes special waste and difficult waste. The latter is a waste which is harmful to humans in the short- or long-term, and potentially harmful to the environment.

Let's take a look at some of the different types of waste, starting with toxic waste.

WASTE
• toxic
• non-toxic
• degradable
• non-degradable
• recyclable
• non-recyclable

DEFINITION
Toxic waste: a poisonous waste that can be harmful or fatal.

EXAMPLES

◆ Toxic waste

Toxic waste is produced in vast quantities and can do great harm to both humans and the environment. The sources of this waste include wood preserving, solvents and degreasing operations, and pesticides.

Examples include cyanide, which is acutely toxic and can kill even in small doses, heavy metals such as lead and mercury, which are used in batteries and are both extremely poisonous to the nervous system, and cadmium, asbestos, and polychlorinated biphenyls (PCBs).

◆ Non-toxic waste

Non-toxic waste is neither fatal nor harmful. Examples include cotton and paper.

◆ Degradable waste

Degradable waste is a toxic or non-toxic waste which can be decomposed. Two common types are biodegradable and photodegradable. Biodegradable waste is decomposed by living matter, an example of which is food, and photodegradable is decomposed by sunlight, an example of which is a plastic bag.

◆ Non-degradable waste

Non-degradable waste is a toxic or non-toxic waste which cannot be decomposed, examples of which are metallic wastes. It is important to realize that nearly all waste degrades by some amount if exposed to the environment.

◆ Recyclable waste

Recycling is the collection and separation of materials from waste and then the subsequent processing to manufacture a product. The UK industry recycles large amounts of waste including:

Bottle recycling plant

- paper and board;
- iron and steel;
- aluminium;
- glass;
- plastics;
- solvents;
- waste oil;
- textiles;
- rubber.

◆ Recycling

Recycling is important to the ecology in various ways:
- it helps to conserve natural resources;
- it cuts down on the volume of waste which has to be transported for disposal in landfill sites and incinerators;
- it reduces environmental damage by indiscriminate dumping.

◆ Non-recyclable waste

Non-recyclable waste is waste which cannot be separated into materials for subsequent processing.

➤ How do we dispose of waste?

Four common methods of disposal of waste are:

- landfill;
- incineration;
- disposal at sea;
- composting.

◆ Landfill

This is the controlled deposit of waste to land in such a way that no pollution or harm results. This method is used for the disposal of 85% of industrial waste.

◆ Incineration

Waste is incinerated in order to sterilize it, and reduce its volume by up to 90%, and its weight by up to 70%; then it is landfilled.

◆ Sea disposal

Waste is dumped at sea.

◆ Composting

Carefully selected organic waste from such sources as the food processing industry is ground down to a particle size of less than about 50mm. The organic waste is then broken down by the action of micro-organisms in the presence of air, that is, biodegraded.

➤ What is a by-product?

A by-product is defined as a secondary or incidental product of a production process. By-products are sold or re-used by the manufacturer and they should not be confused with waste which is disposed of or recycled.

By-products of one production process may well be the raw materials for another production process.

Let's look at some examples of this.

EXAMPLES

Ash is the by-product of burning coal and is sold as one of the raw materials for manufacturing Tarmacadam. Offal is the by-product of the meat industry and is used as one of the raw materials for manufacturing pet food.

This task is carried out in conjunction with tasks 17, 19, and 20

WASTE
• toxic
• non-toxic
• degradable
• non-degradable
• recyclable
• non-recyclable

TASK 18

An investigation into the environmental impact of waste associated with production processes

To carry out this investigation, use all the resources available to you.

Continue with the two contrasting products that you selected in task 17.

Describe the environmental impact of any waste generated by the production processes used to manufacture your products.

Produce a brief report on the environmental impact of any waste generated by the production processes used to manufacture your two selected products.

Keep the brief report in your portfolio.

1.16 ENERGY SOURCES

Energy sources are classified as finite and infinite energy sources.

DEFINITION
Non-renewable energy source: cannot be replenished once used.

DEFINITION
Renewable energy source: can be replenished and so used again and again

FINITE ENERGY SOURCE
A non-renewable source of energy such as fossil fuels which include oil, coal, and gas.

Fossil fuels
Sources of energy formed from the fossilized remains of plant material in the case of oil and coal, and marine plankton in the case of gas.

INFINITE ENERGY SOURCE
A renewable source of energy such as solar radiation, atmospheric winds, geothermal energy, thermal and chemical energy from biomass, and mechanical energy from falling water.

Geothermal
Geothermal energy is thermal energy stored in the rocks and fluids within the earth.

Biomass
Biomass energy is thermal or chemical energy derived from items such as wood, crops, and animal wastes.

Let's begin by looking at the environmental impact of finite energy sources.

➤ Finite energy sources

The burning of fossil fuels causes three major environmental problems: namely air pollution, acid rain, and the greenhouse effect.

◆ Air pollution

We met air pollution as a result of burning fossil fuels in the previous section. Emission of gases such as sulphur dioxide, nitrogen oxides, and carbon monoxide, together with particles affect the whole environment. Humans and animals alike may suffer respiratory problems and vegetation may be damaged.

◆ Acid rain

Sulphur dioxide and nitrogen oxides emitted into the atmosphere during the combustion of fossil fuels can travel from a short distance up to a few hundred kilometres depending upon the height of the chimney stack and the wind speed.

By a complex process involving oxidation, the gases can be transformed into acid compounds of sulphur and nitrogen. These acid compounds, known as acid rain, can fall to the ground in dry weather (dry deposition) or in rain or snow (wet deposition) and in both cases are absorbed by trees, plants, soil and water. Acid rain deposition has had adverse effects on the environment. In affected areas it inhibits plant nutrition, restricts the range of plants, and damages the fish, bird and mammalian populations. It also causes damage to buildings, particularly those made of stone, concrete, and metal.

◆ Greenhouse effect

The atmosphere contains gases known as the greenhouse gases: the main ones are carbon dioxide, methane, ozone, and nitrous oxide. These gases behave, in effect, like glass in a greenhouse: they allow solar radiation to pass through and heat the earth's surface which in turn heats the atmosphere by convection and also radiates heat, some of which passes through the greenhouse gases and some of which is re-radiated back to earth. Without this effect, earth would be virtually uninhabitable.

The combustion of fossil fuels emits gases such as sulphur dioxide, nitrogen oxides, and carbon dioxide. Some of these gases, and in particular carbon dioxide, are greenhouse gases and by altering the level of atmospheric concentration of these greenhouse gases, the global climate may be affected.

This is more commonly known as global warming. It is estimated that if there are no controls put on greenhouse gas emissions, the average global temperature will rise at about 0.5°C per decade with the possibility of significant climate changes. This could lead to the flooding of low-lying land due to the increase in water volume brought about by the melting of ice caps and glaciers.

Let's take a closer look at the environmental impact of fossil fuels, starting with coal, and then oil and gas.

COAL	
BENEFITS TO THE ENVIRONMENT	DETRIMENTS TO THE ENVIRONMENT

Readily available (at present)
 	
Employment
 	
Economic wealth
 	
Major feedstock for the chemical industry
 	
Used to generate power
 	
In 1991, 78% of the UK electricity was from coal
	Combustion of coal:
	   reduces air quality
	   produces acid rain
	   increases greenhouse gases
	
	Non-renewable
	
	Waste tips are unsightly
	
	Waste tips can contaminate water resources with acid run-off
	
	Every year over five million tonnes of minestone spoil is dumped at sea and on beaches
	
	Mining of coal causes land subsidences

OIL	
BENEFITS TO THE ENVIRONMENT	DETRIMENTS TO THE ENVIRONMENT

Readily available (at present)
 	
Employment
 	
Economic wealth
 	
Major feedstock for the chemical industry
 	
Petroleum products include bitumen, naphtha, propane, butane, white spirit
 	
Used to generate power
 	
In 1991, 11% of the UK electricity generation was from oil
	Combustion of oil:
	   reduces air quality
	   produces acid rain
	   increases greenhouse gases
	
	Non-renewable
	
	Environmental pollution from oil spillages which kills marine life and birds, and the plankton that forms the basis of all food relationships in the sea

GAS	
BENEFITS TO THE ENVIRONMENT	DETRIMENTS TO THE ENVIRONMENT

Readily available (at present)
 	
Employment
 	
Economic wealth
 	
Used to generate power
 	
By-products from gas production, such as liquid hydrocarbons, are used in chemical production
 	
Gas is cleaner and more convenient than coal
	Combustion of gas to a lesser extent than coal or oil:
	   reduces air quality
	   produces acid rain
	   increases greenhouse gases
	
	Non-renewable
	
	Methane is a greenhouse gas
	
	Disturbances occurring in gas fields can affect local residents

Let's also consider nuclear power as a finite source.

NUCLEAR	
BENEFITS TO THE ENVIRONMENT	DETRIMENTS TO THE ENVIRONMENT
Produces much less greenhouse gas than conventional power sources Readily available	Waste disposal of radioactive materials Radioactive emissions from nuclear power sources Radioactive contaminations through accidents

Let's now look at the environmental impact of infinite energy sources.

➤ Infinite energy sources

Infinite energy sources can be considered to be virtually inexhaustible; they include solar power, water power, wind power, geothermal and biomass power.

These types of energy source are developed and used at locations where the resource is available; they generally have large space requirements. The initial setting-up costs are high, but the daily operating costs are low and not greatly affected by the conventional energy prices.

Let's consider the environmental impact of the infinite energy sources, starting with solar power.

SOLAR POWER	
BENEFITS TO THE ENVIRONMENT	DETRIMENTS TO THE ENVIRONMENT
Renewable Silent in operation Safe in operation Clean No air pollution	Expensive method of generating electricity (at the moment) Large areas of land are required for a solar power plant Solar power, at present, is generally only used for heating buildings and water; very little is used for generating electricity The intensity of the sun varies according to the weather and the season, and is only available in the daytime, which means that it has to be stored until it is needed Unfortunately it is needed mostly at night and in the winter when it is least available Solar panels are expensive because there is very little mass production of solar energy units

WATER POWER	
BENEFITS TO THE ENVIRONMENT	DETRIMENTS TO THE ENVIRONMENT
(i) Hydroelectricity Renewable No air pollution Produces cheap, constantly available energy Clean In 1991, 2% of the UK electricity generation was from hydropower	Expensive to install Ecological damage because dams are usually needed which means a lot of land is flooded People often have to sell their homes Silt can build up which reduces the efficiency of the hydro-electric plant
(ii) Wave power Renewable No air pollution Waves are strongest in the winter when required most Clean Large number of sites available around the UK	Visual pollution Salt water is corrosive Wave power installations have to be extremely strong to withstand storms
(iii) Tidal power Renewable No air pollution Clean	Expensive to install Barrages built across bays and estuaries may alter the movement of the tides which could cause ecological damage Barrages may also affect coastal sediments and the way that sewage is dispersed into the sea

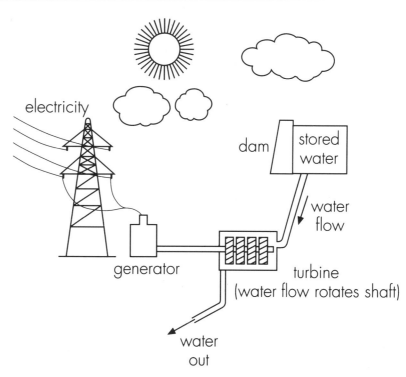

Hydroelectric power station

WIND POWER	
BENEFITS TO THE ENVIRONMENT	DETRIMENTS TO THE ENVIRONMENT
Renewable Clean Wind power is strongest in the winter when it is most needed	Large wind turbines are very noisy Large areas of land are required for a wind turbine plant Visual pollution

GEOTHERMAL POWER	
BENEFITS TO THE ENVIRONMENT	DETRIMENTS TO THE ENVIRONMENT
Renewable Cheaper than conventional power	Produces gas or liquid wastes such as carbon dioxide and hydrogen sulphide, and many poisons such as mercury and arsenic Removing underground rock could cause subsidence or earth tremors Unknown long-term effects of removing the heat from the earth's core

BIOMASS POWER	
BENEFITS TO THE ENVIRONMENT	DETRIMENTS TO THE ENVIRONMENT
(i) Biogas Renewable Rich fertiliser is also produced Relatively cheap power Calorific values compare with conventional fuels	Carbon dioxide is released when organic matter is burnt Contains up to 60% methane
(ii) Burning Renewable Straw and rubbish both provide relatively low levels of energy Disposes of waste	Areas of land are stripped by trees and there is possible soil erosion Burning wood and waste releases carbon dioxide into the atmosphere which adds to the greenhouse effect

This task is carried out in conjunction with tasks 17, 18, and 20

TASK 19

An investigation into the environmental impact of the different energy sources used in production processes

To carry out this investigation, use all the resources available to you.

Continue with the two contrasting products that you selected in task 17.

Describe the environmental impact of the different energy sources used in the production processes to manufacture your two selected products.

Produce a brief report on the environmental impact of the different energy sources used in the production processes to manufacture your two selected products.

Make any necessary assumptions, such as electricity being generated by a coal-fired power station.

Keep the brief report in your portfolio.

ENERGY SOURCES

- finite
- infinite

ENERGY POLICY

- physical energy savings
- energy audit and monitoring
- energy awareness
- energy efficiency

1.17 ENERGY POLICIES

A well-planned and correctly implemented energy policy will not only maximise the efficient use of energy but will also reduce its detrimental effects on the environment.

Let's start by looking at ways of making physical energy savings.

➤ Physical energy savings

Good insulation of all buildings including:

- cavity wall insulation;
- double glazing;
- draught excluders;
- pipe lagging;
- tank lagging.

Install heating controls including:

- thermostats;
- timers.

➤ Energy audit and monitoring

An energy audit is used to control the energy costs of a manufacturing organization by identifying energy wastes. The energy audit is set up by the energy manager and the chief accountant who analyse the current energy use, examine ways of reducing energy wastes and set targets for improving energy use.

An energy audit of a manufacturing process could include some of the following:

- analysis of current energy uses;
- checking unnecessary running of machines;
- checking heat leakages from the process;
- checking whether energy is being wasted in manufacturing a product of excess quality.

A consequence of a well-run energy audit is that the savings in energy costs could mean the products can be more competitively priced.

➤ Energy awareness

In an organization, everyone must appreciate the ways in which energy is used and the ways in which it is wasted. Awareness is best created at the point of use by using posters. For example:

- 'Switch off when you leave';
- 'Don't waste steam, report leaks';
- 'Close factory doors, keep warmth in'.

Other ways of promoting energy awareness are through newsletters, information sheets, and adopting an energy programme logo.

➤ Energy efficiency

There are quite a number of different technological developments available which will help reduce the amount of energy used in manufacturing organizations.

Traditionally, power stations have discarded hot water into rivers and cooling towers. A combined heat and power station pipes the hot water to the local factories and homes. This increases the efficiency of the power station from about 35% to 80%.

Waste materials from a production process can be re-used or recycled. The re-using of waste materials is an efficient method of reducing manufacturing costs and conserving the natural resources. It also has the benefit of reducing the amount of waste that has to be transported for disposal. Recycling is different from re-using in that it involves the collection and separation of materials from the waste which can then be used in the manufacture of products.

The energy efficiency of all machinery and equipment is very important. Regular maintenance and replacement of the machinery is required in order to keep them running efficiently.

The installation of energy-saving fluorescent lamps is worthwhile: they consume about one-fifth of the power of conventional lamps.

This task is carried out in conjunction with tasks 17, 18, and 19

ENERGY POLICY

- physical
- energy audit and monitoring
- energy awareness
- energy efficiency

TASK 20

An investigation into the benefits of an organizational energy policy

To carry out this investigation, use all the resources available to you.

Identify and describe the benefits of an organizational energy policy as applied to a manufacturing process.

Produce a brief report on the benefits of an organizational energy policy as applied to a manufacturing process.

Bring together the brief reports in tasks 17–20 inclusive to produce a report titled: 'Identify the environmental effects of production processes'.

Keep the final report in your portfolio.

2 · WORKING WITH A DESIGN BRIEF

2.1 DEVELOPMENT OF A FINAL DESIGN PROPOSAL

The design process begins with identifying a consumer need and then producing a design brief from it. Once this is established, the next stage is to identify the key design features and production process constraints from the design brief, and then to generate design proposals to meet them. The design proposals are then assessed for feasibility and a final design solution is selected.

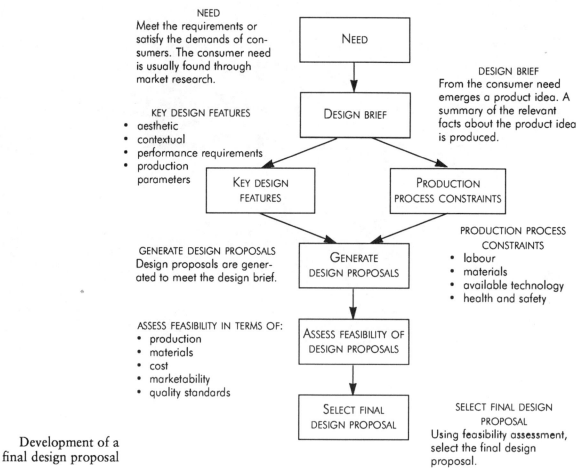

NEED
Meet the requirements or satisfy the demands of consumers. The consumer need is usually found through market research.

KEY DESIGN FEATURES
• aesthetic
• contextual
• performance requirements
• production parameters

DESIGN BRIEF
From the consumer need emerges a product idea. A summary of the relevant facts about the product idea is produced.

PRODUCTION PROCESS CONSTRAINTS
• labour
• materials
• available technology
• health and safety

GENERATE DESIGN PROPOSALS
Design proposals are gener- ated to meet the design brief.

ASSESS FEASIBILITY IN TERMS OF:
• production
• materials
• cost
• marketability
• quality standards

SELECT FINAL DESIGN PROPOSAL
Using feasibility assessment, select the final design proposal.

Development of a final design proposal

2.2   DESIGN BRIEF

A design brief is a description of the client's needs which contains any of the following:

- client's requirements;
- functional details;
- projected cost per item;
- market details;
- any special quality requirements.

A design brief need not necessarily ask for the origination of a new product but may instead ask for the restyling of an existing product or changing of a product to allow for a more cost-effective production.

Let's take a look at an example of a design brief placed by an electrical goods retail outlet into the design section of a medium-sized electrical goods manufacturing company.

➤ Example of a design brief

A recently published survey shows that people who work early or late shifts, women and children, the elderly, walkers and joggers are all at risk from street crime. It is recommended that these 'at risk' groups carry a device that will draw attention to both themselves and the attacker, providing a deterrent whilst summoning assistance.

Design a personal security alarm for 'at risk' groups that will produce a loud, piercing alarm when activated. The device must be lightweight, comfortable to hold in the hand and easy to operate. It must also be compact and tough, functionally reliable and powered by a primary energy source such as a battery or gas cartridge. And, finally, it must cost no more than £10.

This is a stand-alone task and will not form part of any other report

DESIGN BRIEFS

- client requirements
- functional details
- projected cost per item
- market details
- special quality requirements

TASK 21

An investigation into the role of design briefs

To carry out this investigation, use all the resources available to you – such as the library, magazines, journals, teachers, tutors, local companies, and 'yellow pages'.

Investigate the role of design briefs in the world of manufacturing. Who originates them? How common are they? What forms do they take?

Produce a brief report summarizing the role of design briefs in the world of manufacturing.

Keep the brief report in your portfolio

2.3 KEY DESIGN FEATURES

The key design features are identified from a given design brief and agreed with the client. Let's begin by looking at the aesthetic features.

➤ Aesthetic

Aesthetic features are qualities which make a design attractive to look at or pleasing to use. The three main qualities are:

- quality of unity

  Achieved by the simplicity of the shape of the design and the proportion of all the parts which make up the design

- quality of attraction

  Achieved by emphasizing size or colour, or using contrasts to accentuate differences

- quality of balance

  Achieved by forming all the parts of the design together to give a sense of equilibrium, ie stability.

Our senses are used to judge if the above qualities have been achieved in a product:

    touch: a steering wheel or tool handle
    sight: a dress or a pair of trainers
    smell: perfume or an air freshener
    hearing: a pair of loudspeakers or a musical keyboard
    taste: a chocolate bar or a cake.

➤ Contextual

Relevance of the design, ie where and how it will be used.
  For example, to be used:

- by young children;
- by individuals with motor problems;
- by many different operators;
- out-of-doors;
- in a damp environment such as a kitchen.

➤ Performance

The way in which the product is to work. Typical performance characteristics and requirements include:

- speed of operation;
- forces and stresses;
- voltages and currents;
- product life expectancy;
- failure rate;
- strains and deflections;
- acceleration;
- duration;
- working temperature range;
- reliability;
- accuracy of specified parameters;
- output power.

PRODUCTION PARAMETERS

- size
- weight
- cost
- quantity
- time
- quality standards

➤ Production parameters

◆ Size

Reduction in the size of a product may reduce the cost of the product because of the reduction in material volume used. However more exact manufacturing processes may be required and these may increase the cost of the product.

◆ Weight

If the weight of the product has to be within a specified limit, then only certain materials can be used and this can affect the type of production processes used.

◆ Cost

Cost influences most of the decisions associated with designing for production. The two major costs involved are i) its development and ii) the actual cost of the product.

A designer needs to know these costs so that selections between different materials and production processes can be made and thus obtain the lowest product cost. Costs are also directly related to the quantity of products manufactured.

◆ Quantity

The quantity of products to be manufactured will dictate the scale of production, ie continuous flow, repetitive batch, single item, or small batch.

◆ Time

The length of time taken to manufacture a product is directly related to its cost. The longer it takes to manufacture a product, in general the more energy has been provided by the operator and machinery, and hence the more expensive the product is to manufacture.

◆ Quality standards

DEFINITION
Tolerances: total amount of variation permitted for the size of a dimension.
For example, a length of 20 ± 0.1 cm means that the length can lie between 19.9 cm and 20.1 cm.

DEFINITION
Specifications: properties and performance of the product required by the client.
For example: materials, dimensions, and composition.

The quality standards of the product are specified by the client. Quality must be designed and built into a product. To assure that the product is of adequate quality, the following should be specified:

- tolerances;
- material specifications;
- machine finish;
- performance requirements.

Tolerances and specifications that are too rigid make the cost of manufacture expensive. A balance must be achieved between the quality and the cost of the product. Suppose that we have a product with the minimum acceptable quality. If the quality of this product is increased, then there will be a corresponding increase in its cost. Provided that the value of the product is increased by more than the increase in costs incurred by quality improvement, then there is a profitable investment in quality.

Let's take the example of the design brief for a personal security alarm that we saw earlier and agree the key design features with the client.

KEY DESIGN FEATURES

◆ *aesthetic*
- comfortable to hold in the hand
- compact and tough
- red, black and white colour options
- lightweight

◆ *performance*
- powered by a primary energy source such as direct current (DC) batteries or gas cartridges
- must produce a loud, piercing signal
- easy to operate
- functionally reliable

◆ *contextual*
- to be used by people who work early or late shifts, women and children, the elderly, walkers and joggers
- to be used mainly out-of-doors

◆ *production parameters*
- costs – no more than £10
- quantity – 1 000 initially
- time – required in three months
- quality standards – primary energy source must last for a minimum of five minutes under continuous use
- casing of alarm must be made from a plastic material with a high quality finish

---

This task is carried out in conjunction with tasks 23, 24, 25, and 26

TASK 22

An investigation into agreeing with a client the key design features from a design brief

To carry out this investigation, use all the resources available to you – such as the library, magazines, journals, teachers, tutors, local companies, and 'yellow pages'.

Read the following scenarios carefully; they each contain two design briefs.

1. You are a member of a design consultancy who has been approached by a local night-club owner to design a bar stool which will be used by customers who wish to sit at one of the two bars. The stools must be sturdy and comfortable and need not necessarily be made from wood. He will require 20 bar stools to be manufactured.

The owner also wants a 45cm x 45cm carpet tile designed which is hard-wearing, stain resistant, easy to replace and attractive. It is proposed that 3 000 square metres will be covered.

2. You are a member of a design consultancy who has been approached by a local confectioner to design a wedding cake which is to be tiered and to any shape. He has a chain of ten shops in the area and each cake will be produced to customer's order.

He also wants a scone to be designed which is primarily aimed at the health-conscious family. The planned production for the shops is 400 per day.

---

3. You are a member of a design consultancy who has been approached by a nationwide fashion retailer to design a unisex hat (cap) which is for wearing at outdoor events and to be both waterproof and wind-proof, and *not* made of denim. She has 130 shops nationwide and expects 1 600 to be produced each week for three months.

She also wants a unisex scarf designed which is to be formal wear and made of an artificial material. She expects to sell 250 per week nationwide all year round.

---

4. You are a member of a design consultancy who has been approached by a major shoe retailer who wants a unisex sandal designed which is comfortable, hard-wearing and *not* made of leather. She expects to sell 2 000 pairs of sandals per week throughout her chain of 120 shops during the summer months.

She also wants a polish dispenser designed which will dispense polish in liquid form, be non-drip and give an indication to the user of how much polish is remaining. She expects to sell 500 dispensers per week, nationwide, all year round.

---

5. You are a member of a design consultancy who has been approached by a local brewery to design a carton which will contain one-third of a litre of a mixed 'soft' drink. The brewery is to call this new drink 'heavenly juices' and will distribute it to 300 public houses in the region and expects 5 000 cartons to be produced each week.

The brewery also wants a label designed for a new UK wine that they are marketing and distributing. It is a medium-dry white wine and they expect to market the wine nationwide, producing 7 500 bottles each week.

---

6. You are a member of a design consultancy who has been approached by a small local publisher to design the cover for a new book. The book is by a local author who has researched the historical development of his area and put it into words and photographs. The publisher expects 2 000 copies to be produced each week for three months.

The publisher also wants to mark the 50th anniversary of her business with a specially designed bookmark. She expects to distribute 1 000 of them.

Select any two design briefs from the scenarios. For example the wedding cake from scenario 2 and the bookmark from scenario 6.

Make your selections carefully – you will be asked to 'manufacture' one of the items later.

Identify what you feel are the key design features from your two design briefs.

Agree the key design features from both design briefs with your teacher or tutor, or local manufacturer.

Summarize and record the key design features from both design briefs.

Keep the key design features in your portfolio.

KEY DESIGN FEATURES

- aesthetic
- contextual
- performance
- production parameters

PRODUCTION PROCESS CONSTRAINTS

- labour
- available materials
- available technology
- health and safety
- quality standards

LABOUR

- skill
- personnel

2.4 PRODUCTION PROCESS CONSTRAINTS

Production process constraints are identified from the design brief. Let's begin by looking at the labour production constraint.

➤ Labour

◆ Skill

New products might use different types of material which each may need different types of production process.

The workforce may need to be trained or re-skilled in these new techniques.

If the skills or production processes are not available then the designer may alter the product accordingly.

◆ Personnel

The scale of production of a new or re-designed product will influence the size of the workforce.

More workers may need to be employed or sub-contracted.

Conversely a new design may reduce the size of the workforce.

The design of the product may be influenced by the existing size of the company workforce.

➤ Available materials

Ideally a product should be designed or re-designed to use materials which are readily available, provided the quality of the product can be maintained. Materials which are not readily available might perhaps be relatively more expensive, might not be received on time, and might hold up production causing potential losses to the manufacturer.

Readily available materials will help to keep the cost of the manufactured product at a competitive level.

The selected materials for the product have a major influence on the production process that is used to manufacture the product.

➤ Available technology

A new product should ideally be designed or re-designed using the most up-to-date technology available as long it is proven and well-documented and can provide a quality product that is more economical than the competitors.

However, using this technology in the new product may require the installation of a different type of production process, and the subsequent re-skilling and training of the workforce in the new techniques. Manufacturing companies have to keep up with all the changes in technology if they are to stay competitive in the world of manufacturing.

Here are some examples of current available technologies:

EXAMPLES

- the manufacture of squash rackets has changed in the last few years: rackets were made of wood, then aluminium, then graphite. Today they are being designed and manufactured in one-piece constructions using kevlar/graphite/epoxy composite frames;
- trainers are now being designed and manufactured using super lightweight, multi-polymer composite materials;
- soft-drink and beer cans are now made out of aluminium using advanced manufacturing techniques.

➤ Health and safety

Production processes could involve any of the following:

- use of toxic or hazardous materials;
- handling and disposal of waste products;
- changes in humidity;
- smell;
- pollution;
- electrical shock;
- noise;
- radiation;
- magnetic properties;
- radio frequency (RF) interference.

➤ Quality standards

The constraint is whether the production process is capable of meeting the quality standards specified by the client.

In particular, it is important to examine whether it can meet the following:

- specified tolerances;
- required material specification;
- specified machine finishes;
- specified performance requirements.

Let's take the example of the personal security alarm and identify the possible production process constraints from the design brief.

PRODUCTION PROCESS CONSTRAINTS

◆ **Available technology**

The manufacturing company produces a range of consumer goods but does not have the necessary production processes to manufacture the plastic casings. It will either have to buy the casings in or install a new production process to manufacture them. The personal security alarm will be designed with current available technology, and all components and parts will be bought-in and assembled in the plant.

◆ **Labour**

If a new production process is installed to manufacture the plastic casings, then the workforce will need re-skilling and training in the new techniques. The workforce may decrease if it is a new product which is replacing an existing one, and being manufactured using current available technologies. Alternatively, the workforce may increase if the new product does not replace an existing one.

◆ **Materials**

The client wants the personal security alarm to be manufactured using a plastic casing. Plastics are polymers that can easily be moulded into the required shape by the application of heat, which is retained when it has cooled. A thermoplastic can be remoulded by reheating it, whereas a thermoset retains its moulded shape when it is reheated. Plastics can be manufactured in any desired colour or finish.
The four main thermoplastic polymers are:

- polyethylene (polythene);
- polystyrene;
- polypropylene;
- polyvinyl chloride (PVC).

Thermoset polymers include:

- phenolics;
- epoxides.
- amino-formaldehyde;

There is such a large number of different types of plastic, all with their own characteristics and properties, that the selection of the right plastic for a particular application is best done using up-to-date information from the material suppliers.

◆ **Health and safety**

If any chemicals are used in the production process, then extreme caution will have to be exercised when handling them.
    Ventilation will have to be provided against vapours that may result from the production process.
    Appropriate safety clothing will need to be provided.
    Very little waste disposal is anticipated since most material can be re-used or recycled.

This task is carried out in conjunction with tasks 22, 24, 25, and 26

PRODUCTION PROCESS
CONSTRAINTS

- labour
- materials
- available technology
- health and safety

TASK 23

An investigation into the identification of production process constraints from a design brief

To carry out this investigation, use all the resources available to you.

Identify the production process constraints from your two selected design briefs in task 22.

Summarize and record the production process constraints from both design briefs.

Keep the records in your portfolio.

2.5 GENERATING POTENTIAL DESIGN PROPOSALS

Let's attempt to devise possible proposals for the design brief of the personal security alarm. The design brief states that we can use a primary energy source such as batteries or gas cartridges.

Let's consider batteries first.

➤ DC battery-powered personal security alarm

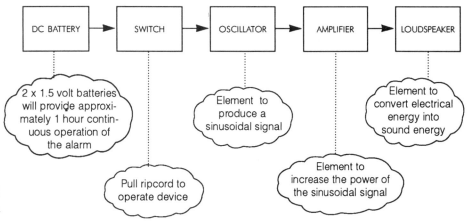

DC battery-powered personal security alarm

The personal security alarm operates like this:

Pulling the ripcord removes a plug from the device and this allows current supplied from the DC battery to enable the oscillator to begin producing sinusoidal signals. These signals are amplified and fed through a loudspeaker which emits a loud piercing signal.

When the plug is replaced in the device, the oscillator is disabled and the loud piercing signal stops.

Now let's generate a proposal for the gas cartridge personal security alarm.

➤ Gas cartridge-powered personal security alarm

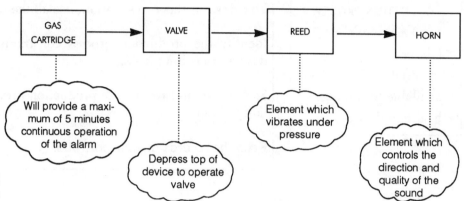

Gas cartridge-powered personal security alarm

The horn will be the actual shape of the personal security alarm.

Operation of the device is as follows:

When the top of the alarm is depressed, a valve is operated which allows pressure from the gas cartridge to vibrate the reed rapidly causing a loud piercing signal to be emitted from the horn.

We have now generated two potential proposals for the personal security alarm. One is battery-powered and the other is gas-powered. We have seen the proposals in the form of block diagrams – but what will they actually look like?

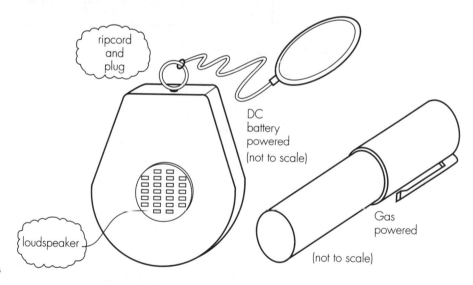

Personal security alarms

On the opposite page we will compare the key design features and then the production process constraints of the two design proposals for the personal security alarm.

DESIGN PROPOSALS FOR THE SECURITY ALARM		
KEY DESIGN FEATURES	DC BATTERY-POWERED	GAS CARTRIDGE-POWERED
(i) Size	Fits comfortably in the hand More compact than gas-powered device	Fits comfortably in the hand
(ii) Lightweight	Approximately 70g with batteries	Approximately 100g with cartridge
(iii) Functional reliability	Uses modern technology Needs far more components than gas-powered device A battery level indicator has *not* been designed into the device	Well-proven technology Will need periodic maintenance
(iv) Easy to operate	The device is activated by pulling the 'rip-cord' which removes the plug from the device and the alarm is sounded Re-insertion of the plug stops the alarm Perhaps more difficult to activate than the gas-powered device A slide switch to activate the device could be added if the client so wished	The device is activated by depressing its top firmly
(v) Audible & continuous alarm	Alarm is loud and continuous for approximately 60 minutes (if required)	Alarm is very loud and continuous for approximately 5 minutes
(vi) Robust (strong & sturdy)	Will be manufactured in tough plastic	Will be manufactured in tough plastic but the shape is more susceptible to damage than the DC-powered device
(vii) Powered by primary energy source	3 volts DC supply from 2 x 1.5 volt batteries Probably more readily available than gas cartridges	Uses gas cartridge

DESIGN PROPOSALS FOR THE SECURITY ALARM		
PRODUCTION PROCESS CONSTRAINTS	DC BATTERY-POWERED	GAS CARTRIDGE-POWERED
(i) Re-skilling of work-force	Re-skilling will be needed if injection moulding process is installed (Injection moulding involves melting the thermoplastic polymer and forcing it into a mould, and then allowing it to set hard)	Re-skilling will be needed if extrusion process is installed (Extrusion process involves forcing a thermoplastic polymer through a die)
(ii) Materials	Polypropylene has been selected because it is a rigid and tough plastic, and is readily available	Polypropylene has been selected because it is a rigid and tough plastic, and is readily available
(iii) Available technology	Injection moulding process to manufacture plastic casings Integrated circuitry	Extrusion process Valve and reed
(iv) Health & safety	Appropriate safety clothing will be needed Adequate ventilation will be required	Appropriate safety clothing will be needed Adequate ventilation will be required

This task is carried out in conjunction with tasks 22, 23, 25, and 26

TASK 24

Generating proposals to meet the design brief

To carry out this investigation use all the resources available to you.

Generate at least two proposals for each of your two selected design briefs in task 22.

Show how each design proposal meets the requirements of the design brief.

Summarize and record the proposals generated to meet the design briefs.

Keep the proposals in your portfolio.

2.6 ASSESSING THE FEASIBILITY OF DESIGN PROPOSALS

The feasibility of each design proposal is assessed in terms of:

- production;
- materials;
- cost;
- marketability;
- quality standards.

Let's assess the feasibility of the two design proposals for the personal security alarm.

FEASIBILITY OF THE DESIGN PROPOSALS		
ASSESSED IN TERMS OF	DC BATTERY-POWERED	GAS CARTRIDGE-POWERED
(i) Production	Plastic casing can be manufactured using injection moulding process Method of assembly must be carefully considered at the design stage because it is an expensive cost element Assembly time can be reduced by using 'snap together' parts Moulding for casing requires battery compartment, slots to hold printed circuit board (PCB) and compartment for miniature type loudspeaker It is feasible to produce	Plastic casing can be manufactured using extrusion process Shaped to fit gas cartridge It is feasible to produce
(ii) Materials	Plastic casing made of polypropylene Printed circuit board (PCB) containing electronic circuitry including oscillator and amplifier Type of technology is chosen to keep the power consumption to a minimum in order that a 3-volt DC battery can be used Materials are feasible to obtain	Plastic casing made of polypropylene Valve Reed Materials are feasible to obtain

FEASIBILITY OF THE DESIGN PROPOSALS		
ASSESSED IN TERMS OF	DC BATTERY-POWERED	GAS CARTRIDGE-POWERED
(iii) Cost	Less than £10 It is feasible to manufacture for less than £10	Less than £10 It is feasible to manufacture for less than £10
(iv) Marketability	Walkers and joggers would prefer the DC battery-powered personal security alarm because it is compact, lightweight, and can be strapped to the wrist or attached to a key ring It is feasible to market it for walkers and joggers	The elderly and possibly some women and children would prefer the gas-powered personal security alarm because it is easy to operate and produces a very loud sound It is feasible to market it for the elderly, women and children
(v) Quality standards	The batteries will last for much longer than 5 minutes under continuous use The finish of the personal security alarm will be high quality It is feasible to produce the personal security alarm to the specified quality standards	The gas cartridge will last for five minutes under continuous use The finish of the personal security alarm will be high quality It is feasible to produce the personal security alarm to the specified quality standards

There is an opening in the market for both types of personal security alarm.

This task is carried out in conjunction with tasks 22, 23, 24, and 26

FEASIBILITY ASSESSED
IN TERMS OF:
- production
- materials
- cost
- marketability
- quality standards

TASK 25

Assessing the feasibility of design proposals

Use all the resources available to you – such as the library, magazines, journals, teachers, tutors, local companies, and 'yellow pages'.

Assess the feasibility of each of your design proposals from task 24.

Summarize and record the feasibility assessments of your design proposals generated to meet the two design briefs.

Keep the notes in your portfolio.

This task is carried out in conjunction with tasks 22, 23, 24, and 25

TASK 26

Selecting your design proposal for presentation

Use all the resources available to you.

Select one design proposal from those assessed in task 25 for each design brief.

Show how the feasibility assessment was used to select your final design proposals for the two design briefs.

Record the notes on how the feasibility assessment was used to select your final design proposals for the two design briefs.

Bring together tasks 22–26 inclusive to produce a report titled: 'Originate product proposals from a given design brief'.

Keep the report in your portfolio.

SUPPORT MATERIALS FOR PRESENTATIONS

- notes
- overhead transparencies
- diagrams
- tables
- charts
- graphs
- technical reports
- technical drawings
- pictorial drawings
- models
- photographs
- handouts
- feedback forms

2.7 SUPPORT MATERIALS FOR PRESENTATIONS

Support materials such as notes, overhead transparencies and models are an important part of any design proposal presentation. They allow the presenter to explain the design and production details of the proposal to the clients in an effective and professional manner.

Let's begin by looking at the use of notes as a support material for presentations.

➤ Notes

Write the presentation out in full.

From this make reasonably detailed notes. If you find that the words do not 'flow' when you are presenting your design proposal, then you have the detailed notes to fall back on. Make sure that you feel 'comfortable' with everything that you put in the notes. It is no good writing down notes which you do not fully understand.

Notes should be designed to be a support and not just something to fall back on. Notes should be written on cards in large, clear lettering. If possible use different coloured cards to represent different parts of the presentation, such as the cue for a visual aid. Use highlighter pens to emphasize important points.

➤ Overhead transparencies

This is an attractive method of presenting information and data. Overhead transparencies, commonly called acetate sheets, are used with overhead projectors (OHP); they can be written on using either permanent or non-permanent types of coloured pen; lettering can be applied more professionally using Letraset dry letter transfers. Information can be photocopied on to specially-treated acetate sheets, but be careful – this can infringe copyright control laws.

Diagrams, tables, charts, graphs, and many different types of illustration can be very effectively conveyed to the audience. Good use of colour and shading on the transparency will assist the audience in understanding the information being conveyed.

During the presentation additional information can be added to the transparency to build up a more complex picture in front of the audience; alternatively, a succession of overlays can gradually be laid on top of the original transparency to build up the picture.

➤ Diagrams

Diagrams are an efficient method of presenting graphic information. They use conventions and symbols, and may be pictorial drawings (isometric or oblique) or orthographic drawings.

Diagrams can illustrate complex ideas without excessive details. They should be well-labelled and uncluttered; if you like, the important parts can be highlighted. Insets can be included to show magnified versions of a section of the diagram.

A diagram should be given a title, normally framed in a box, and an identifying number followed by a short description, for example: *Fig 2.3 Typical cooking utensils.*

Let's look at three common types of diagram:

- pictorial diagrams;
- flow diagrams (also called block diagrams);
- circuit diagrams.

◆ Pictorial diagram

Pictorial diagrams are used to describe processes or systems using pictorial symbols and interconnecting lines.

On the opposite page you will find two pictorial diagrams, one of a typical canning process in the food industry and the other of the manufacture of laminated paper tubes.

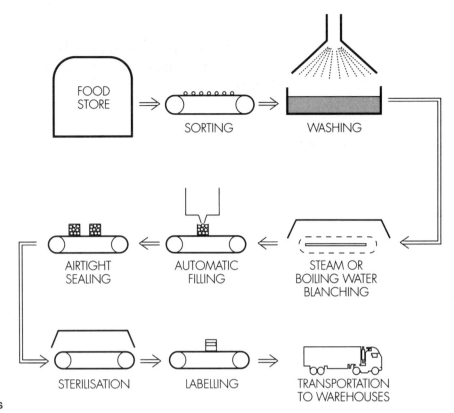

Typical canning process

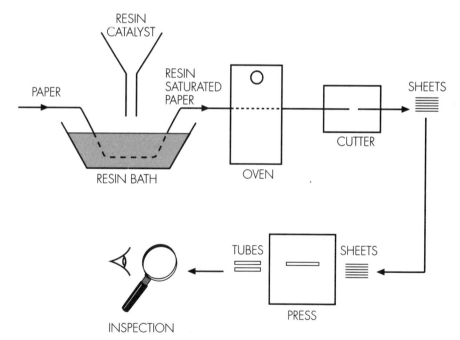

Manufacture of
laminated paper tubes

◆ Flow diagram

Flow diagrams are used to describe processes or systems using boxes, interconnecting lines and arrows. Let's look at the most basic flow diagram.

Basic system

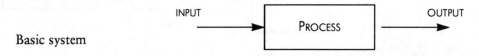

And using this building block, a more complex flow diagram can be built up.

More complex system

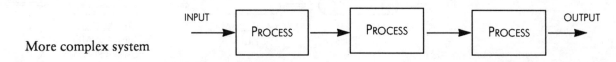

Let's take a look at an example of a flow diagram illustrating the production of wheat flour.

Production of wheat flour

◆ Circuit diagram

Circuit diagrams are used to describe processes or systems using symbols and interconnecting lines.

Let's take a look at the circuit diagram of an amplifier with a gain of 10.

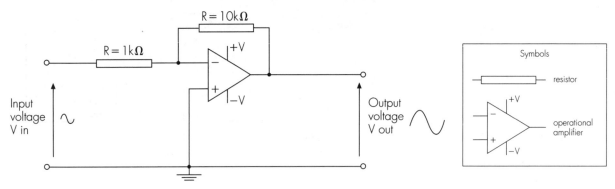

Amplifier with a gain of 10

Here is another example of a circuit diagram, known as a logic circuit.

INPUTS

A

B

A

$\overline{B}$

$A.\overline{B}$

OUTPUT

$F = A.\overline{B} + \overline{A}.B$

B

$\overline{A}$

$\overline{A}.B$

Exclusive OR-gate

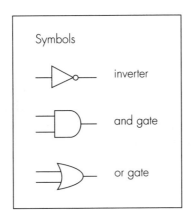

Symbols

inverter

and gate

or gate

A	B	F
0	0	0
0	1	1
1	0	1
1	1	0

Truth table

Now let's have a look at tables.

➤ Tables

Tables are a concise method of displaying information such as statistics, trends and changes. The table should have a title, and each column in the table should have a heading. The information in the table should be organized and presented in a logical manner.

Here are two examples of tables:

AREA	NO OF DESIGNERS
North	600
South	450
East	380
West	520
TOTAL	1950

Distribution of designers
in Dorset, 1996

Work role	Number employed	Percentage (%)
Process operator	410	82
Clerks	50	10
Maintenance staff	30	6
Designers	10	2
Total	500	100

Distribution of workers employed in a medium-sized plastics company

The next support material to investigate is the chart.

➤ Charts

There are many different types of chart and we will look at three of the more common ones, the pie chart, the bar chart, and the flow chart.

We will begin with the pie chart.

◆ Pie chart

The pie chart is so called because it is circular in shape and each segment looks like a slice of pie. It is an effective method of visually conveying the percentage components that make up a given total.

Consider the nutritional information of a sauce mix that is being proposed as a new product for the food market.

Nutritional information	per 100g
Protein	25g
Carbohydrate	58g
Fat	17g
Total	100g

Nutritional information of a proposed sauce mix

We need to convert each component into a percentage component.

In every 100g of the sauce mix there are 25g of protein. Therefore there is: $\dfrac{25\,\text{g}}{100\,\text{g}} \times 100\%$ which is 25%.

The other percentage components can be worked out in the same way and then we can form the following table:

Nutritional information	per 100g	%
Protein	25g	25%
Carbohydrate	58g	58%
Fat	17g	17%
Total	100g	100%

We now have the percentage components that make up the total.

How do we represent these percentage components on a pie chart? Well, first of all we need to be able to relate the percentage component to the angle its corresponding segment makes at the centre of the circle.

The total, ie 100%, is proportional to the number of degrees in a circle, ie 360°.

$$100\% : 360°$$

Divide both sides by 100

$$1\% : 3.6°$$

This means that every 1% of the component is allocated 3.6° as a segment angle.

For example, 25% is allocated: 25 x 3.6° = 90°.

The other two segment angles can be worked out in the same way, and thus we form the following table:

NUTRITIONAL INFORMATION	PER 100g	%	DEGREES
Protein	25g	25	90
Carbohydrate	58g	58	208.8
Fat	17g	17	61.2
TOTAL	100g	100%	360°

We can now produce a pie chart of the nutritional information of a proposed sauce mix.

Pie charts of the nutritional information of a proposed sauce mix

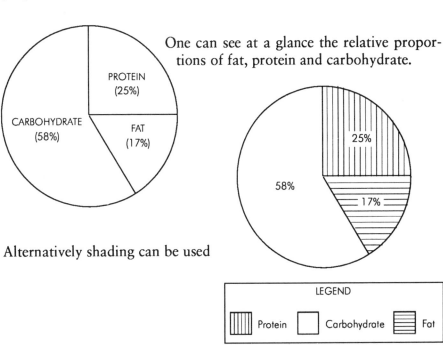

One can see at a glance the relative proportions of fat, protein and carbohydrate.

Alternatively shading can be used

Let's follow that with a look at bar charts.

◆ Bar chart

The bar chart represents data with a series of bars or strips, the width of each bar being unimportant but the length of the bar being proportional to the quantity being represented. The bars can be drawn either horizontally or vertically.

Let's draw a vertical bar for the nutritional information of the proposed sauce mix.

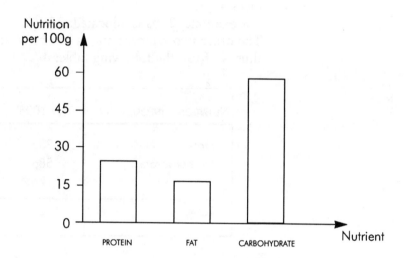

Vertical bar chart for the distribution of nutrients in each 100g of a proposed sauce mix

Another important type of bar chart is a component bar chart. This type of bar chart enables the audience to make comparisons between the sets of components.

As an example, suppose the proposed sauce mix is replacing an existing mix which has the following nutritional information:

NUTRITIONAL INFORMATION	PER 100g
Protein	17g
Carbohydrate	62g
Fat	21g
TOTAL	100g

Nutritional information of existing sauce mix

Let's draw a component bar chart to compare the nutrients of the two sauce mixes per 100g.

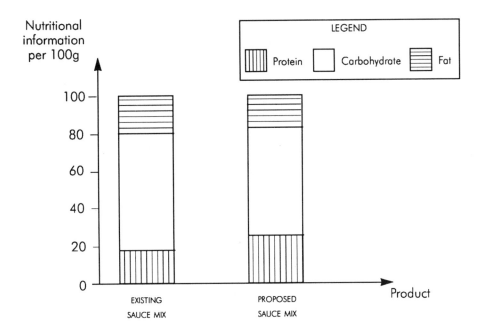

Component bar chart comparing the distribution of nutrients in the existing and proposed sauce mixes

◆ Flow chart

A problem can be made easier to solve by breaking it down into a series of steps known as an ALGORITHM. This can be best achieved by producing a flow chart.

Here are some typical flow chart symbols.

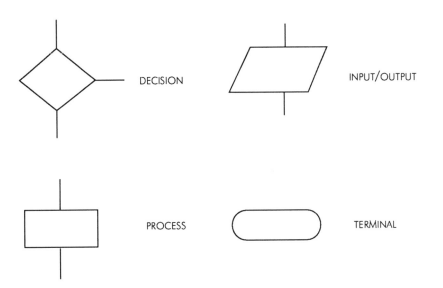

On the next page you will find an example of a simple flow chart, drawn using some of these symbols.

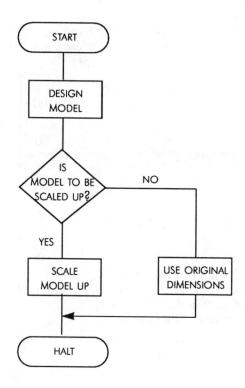

Simple flow chart

➤ Graphs

A graph is a drawing showing a functional relationship between two (or more) variables.

Let's consider the following information regarding the number of designs produced by Inventalot plc between 1993 and 1996 inclusive given in the following table.

YEAR	NUMBER OF DESIGNS
1993	71
1994	31
1995	18
1996	13

Designs produced by
Inventalot plc 1993–96

◆ Single line graph

Let's now produce a single line graph of the information given in the above table.

The single line graph on the opposite page reveals some very interesting points that are not readily available from the table.

It shows that there has been a continual decrease in the number of designs produced from 1993 onwards, and that there was a sharp fall in output between 1993 and 1994, and a more gradual reduction thereafter.

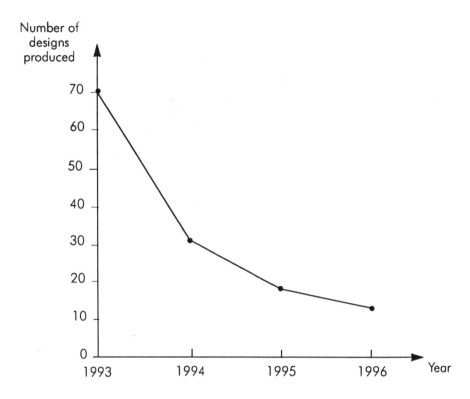

Number of designs produced

Single line graph of the designs produced by Inventalot plc 1993–96

Year

◆ Multiple line graph

Another popular type of graph is a multiple line graph. This type of graph is used to show the comparison of trends to the audience.

As an example, suppose we also have the number of designs produced by Makealot plc between 1993 and 1996 inclusive.

YEAR	NUMBER OF DESIGNS
1993	65
1994	47
1995	16
1996	28

Designs produced by Makealot plc 1993–96

Let's produce a multiple line graph of the designs produced by Inventalot plc and Makealot plc between 1993 and 1996 inclusive.

We have already seen on the single line graph that there had been a continual decrease in the number of designs produced by Inventalot plc from 1993 onwards. Makealot plc also show a decrease in output from 1993 to 1995 but then start to increase their output in 1996. On the next page you will see this comparison as a multiple-line graph.

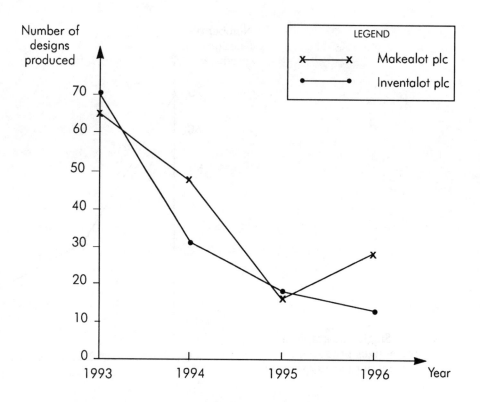

Multiple line graph comparing the number of designs produced by Inventalot plc and Makealot plc between 1993 and 1996

Right, let's now have a look at the technical report and its main points.

➤ Technical reports

The purpose of a report is to record, inform and recommend. A typical report structure is as follows:

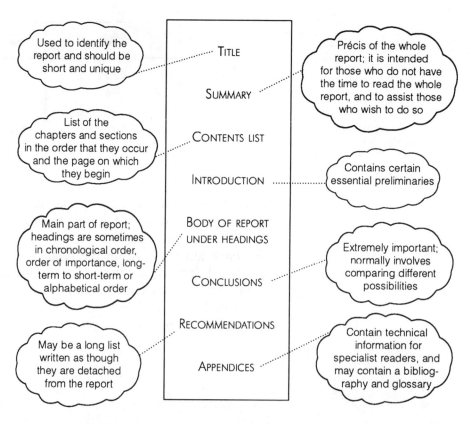

Used to identify the report and should be short and unique

TITLE

Précis of the whole report; it is intended for those who do not have the time to read the whole report, and to assist those who wish to do so

SUMMARY

List of the chapters and sections in the order that they occur and the page on which they begin

CONTENTS LIST

INTRODUCTION

Contains certain essential preliminaries

Main part of report; headings are sometimes in chronological order, order of importance, long-term to short-term or alphabetical order

BODY OF REPORT UNDER HEADINGS

CONCLUSIONS

Extremely important; normally involves comparing different possibilities

RECOMMENDATIONS

May be a long list written as though they are detached from the report

APPENDICES

Contain technical information for specialist readers, and may contain a bibliography and glossary

Structure of a report

➤ Technical drawings

Technical drawings are produced to the recommendations of the British Standards Institution (BSI) who are the UK national standards body. Recommendations include drawing layout, scales, types of line, lettering, methods of orthographic projection, sections, and conventional representations.

Two important types of technical drawing are:

- detail drawings;
- assembly drawings.

Each drawing should have some of the following basic information displayed on it:

- name of company;
- drawing number;
- title;
- date;
- signature of draughtsperson;
- projection symbol (first or third angle);
- unit of measurement (inch or millimetre);
- original scale;
- sheet number and number of sheets;
- issue information.

◆ Detail drawings

A detail drawing defines an individual part or piece for manufacture and includes all the necessary information required to define the part or piece completely. This information can include the following:

- dimensions and tolerances;
- materials;
- scale;
- drawing number;
- description of the system to which the part or piece belongs.

- surface finish;
- number of parts;
- title of the object;
- designer's name;
- date;

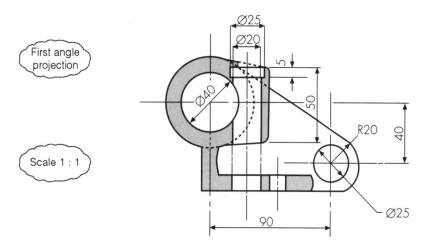

Part of a detail drawing

◆ Assembly drawing

An assembly drawing shows all the component parts and/or sub-assemblies of the product in their assembled form. Assembly drawings can contain the following information:

- title;
- drawing number;
- designer's name;
- quantity per assembly of each part.

- scale;
- date;
- parts list;

Assembly drawings are used to determine the manufacturing schedules.

Let's take a look at an example of an assembly drawing drawn from the detail drawing.

PART NO	TITLE	NO/SET
1	Body	1
2	Hex nut M10	1
3	Plain washer	1
4	Sleeve	1
5	Bolt	1
6	Bush	2

 First angle projection

 Scale 1 : 1

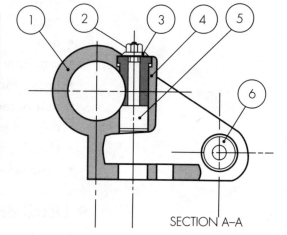

SECTION A–A

Part of an assembly drawing

➤ Pictorial drawings

Pictorial drawings, sometimes called illustrations, are a useful way of showing potential customers what the product actually looks like.

A popular type of pictorial drawing is the exploded drawing which shows how the different parts of the product are assembled.

Let's look at an example of an exploded drawing of the product that we saw earlier in the technical drawings.

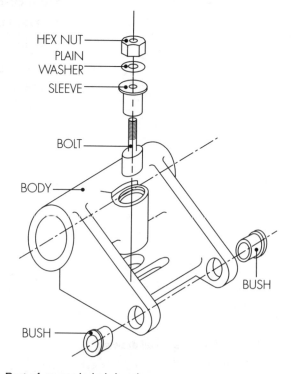

Part of an exploded drawing

116   *WORKING WITH A DESIGN BRIEF*

➤ Models

Models can be static (ie non-working) or working. They can be scaled-up, scaled-down or full scale. A model can be dismantled and re-assembled, and non-essential detail can be omitted. Different parts of the model can be coloured so that they can be picked out easily, and the speed of operation can be scaled appropriately. Models provide a three-dimensional 'sensation' of the design.

➤ Photographs

A photograph represents a 3-D object in two dimensions. Photographs are very useful when different angles of a design are required. A series of photographs can be taken to show a process from start to finish. An important factor is that a photograph can be in colour or black and white.

➤ Handouts

Handouts should be well prepared. If at all possible, they should be word-processed and contain the same terminology as in your presentation. Their uses vary from providing the audience with a prior background knowledge to a summary of your presentation.

➤ Feedback forms

Feedback forms should be designed to be used by the audience at the end of your presentation to give you both quantitative and qualitative judgements (see section 2.11 for definitions) about your design proposal.

All questions on the form should be simple and unambiguous, and designed to provide you with maximum useful feedback information.

This task is carried out in conjunction with tasks 28, 29, 30, 31, and 32

SUPPORT MATERIALS

- notes
- overhead transparencies
- diagrams
- tables
- charts
- graphs
- technical reports
- technical drawings
- pictorial drawings
- models
- photographs
- handouts
- feedback forms

TASK 27

Preparation of appropriate support materials to explain design and production details of selected product proposals

Use all the resources available to you, particularly tasks 22–26 inclusive.

Prepare the appropriate support materials that you would use to explain the design and production details of the two design proposals that you finally selected in task 26. Here are lists of the support materials, design details, and production details you will need.

DESIGN DETAILS	PRODUCTION DETAILS
• aesthetic	• quality of materials
• contextual	• size
• performance	• manufacturing methods
• production parameters	• finish

Preparation for one of your product proposal preparations must be carried out in detail, but the other can be done in outline.

Use the appropriate technical vocabulary when preparing your support materials. Make sure it relates to each product and to the audience to whom you are going to give your presentations. Some commonly used technical words and a brief definition are listed in section 2.9.

Keep a record of all the preparatory work in your portfolio.

PRESENTATION
TECHNIQUES

- verbal
- visual
- written

2.8 PRESENTATION TECHNIQUES

Communication in a clear and concise manner, irrespective of whether it is through the spoken or written word, or using visual aids is extremely important if your presentation is to be effective.

Let's begin by looking at verbal presentation techniques.

➤ Verbal

Practise your presentation until you are satisfied with it.

Concentrate on getting those difficult or commonly mispronounced words right. Try and find alternative words with which you are happy.

When giving your presentation, be enthusiastic and let your voice show that you know your subject.

Vary the tone of your voice to make sure it does not become monotonous. Speak clearly and distinctly. Project your voice but do not shout.

Do not repeat yourself.

Attempt to get the audience on your side.

Do not hide behind your notes or script if you have to read directly from them.

Try to stand reasonably still whilst speaking and make sure you look at the whole audience. Keep your hands still by your side.

Let the audience know that you are the one giving the presentation.

Now let's look at visual presentation techniques.

VISUAL TECHNIQUES

- overhead projector
- whiteboard
- video
- slides
- computer
- flipchart

➤ Visual

Visual techniques such as the use of an overhead projector, slides, or a computer will all enhance your presentation by capturing the audience's attention and helping you in putting your message across.

Carefully check all your visual aids, such as overhead transparencies and slides, well before your presentation. Use them when rehearsing your presentation, even if you cannot project them.

Remember, too many transparencies and/or slides can have an adverse effect on your presentation. Make a limit of four or five of each.

Before you give your presentation, check that all the equipment such as overhead projector and slide projector is functioning, and that you have positioned and adjusted them so that your displays are in focus.

Make sure that you have laid out all your visual aids in the order that you are going to use them. It is very easy to put slides upside down in the carousel!

Let's look at the various visual techniques in a little more detail.

◆ Overhead projector (OHP)

The OHP projects the information on your transparencies on to a white background in such a way that the whole audience should be able to see them clearly. An important feature is that you can easily return to previously-used transparencies if necessary. An OHP helps to present information in a professional manner.

◆ Whiteboard (chalkboard)

This is useful for making short notes, but you do need to be able to write and draw neatly. It is not normally a particularly impressive method of formally presenting information.

◆ Video

A video is an excellent method of presenting information if it is well-produced. Unfortunately, it is very time consuming and expensive to produce a video of the right quality for a presentation. It is better not to use a video than to use a poor-quality one in a presentation.

◆ Slides

Photographic slides are an impressive way of presenting information; however, they are expensive to produce. They help in keeping the audience's interest, and allow the presenter to narrate the presentation around each slide.

◆ Computer

The information displayed on a computer (video display unit) can be projected using a specialist piece of equipment on to a white screen (like the OHP presentation). This allows you to utilise many different types of application programs: databases, spreadsheets, word-processing and computer-aided design (CAD) in your presentation. Don't forget that if you use a computer in your presentation, you can also produce print-outs of your various displays as hand-outs for the audience.

◆ Flip chart

A flip chart is a large pad of white paper placed on an easel. It is not a terribly impressive method of presenting information, but it is useful for interacting with the audience and making short notes. You can prepare short items of information prior to the presentation and use them as necessary. Information can be left displayed as it does not cause such a distraction as the light from an OHP.

And finally, let's look at written presentation techniques.

➤ Written

Written presentation techniques have been covered extensively in section 2.7, and are an integral part of visual techniques, such as overhead transparencies, whiteboard and flipchart.

This task is carried out in conjunction with tasks 27, 29, 30, 31, and 32

PRESENTATION TECHNIQUES
• verbal
• visual
• written

AUDIENCE
• familiar
• client

TASK 28

Using the presentation techniques appropriate for the product type and audience

Use all the resources available to you.

Decide and plan what presentation techniques you are going to use that are appropriate to each of your two product proposals and to the audience in each case.

Discuss with your teacher/tutor what type of audience you will have for each of your two presentations. Perhaps your detailed presentation could be given to a familiar audience and the outline presentation to a client, or vice versa.

Write out in detail the presentation for one of the product proposals and in outline for the other. Remember you decided which one would be detailed in task 27.

Use the appropriate technical vocabulary, related to each product and to the audience to whom you are giving the presentation.

Keep a record of all the work in your portfolio.

2.9   TECHNICAL VOCABULARY

Here are some commonly used technical words and a brief definition which could be helpful for your presentation and support materials.

TECHNICAL WORD	DEFINITION
analysis	detailed examination
appearance	how something looks
assembly	process of putting together
characteristic	distinguishing quality
complexity	intricacy
components	constituent parts
constraints	restrictive conditions
construct	assemble or build
conventional	customary or traditional
criteria	standards by which something can be judged
demand	need or requirement
development	act of growing or developing
effectiveness	capability of producing a result
ergonomics	study of the relationship between workers and their environment

Technical word	Definition
evaluation	assessment or calculation
function	to operate or perform as specified
implementation	putting into action
initiative	commencement or first step
integration	combining or blending
integrity	soundness
maintenance	keeping in proper or good condition
marginal	close to a limit
marketing	business of selling goods
modification	an adjustment or alteration
option	choice
output	the amount produced, usually in a given time
performance	quality of functioning
pollution	contamination normally by harmful or poisonous substances
process	series of operations
prototype	trial model or preliminary version
quality	standard of excellence
reliability	dependability
research	to carry out investigations
specification	detailed description of the construction, materials, and workmanship required to manufacture a product
standard	an accepted or approved example of something against which others are judged or measured
synthesis	putting together
system	a set of related components
tolerance	permissible variation

This task is carried out in conjunction with tasks 27, 28, 30, 31, and 32

TASK 29

Using appropriate technical vocabulary

Use all the resources available to you.

Technical vocabulary appropriate to the product proposal and to the audience should be used in each of your presentations.

Produce a separate list of the technical words and their brief definitions for each of your two presentations.

Keep a record of each list in your portfolio.

2.10 PRESENTING DESIGN PROPOSALS

Let's take a look at the type of information that would be presented for a design proposal. Let's begin with a design proposal for a new food product, a crunchy new-style milk chocolate biscuit. Figures (a) to (i) show various ways of presenting the information.

➤ Chocolate biscuit

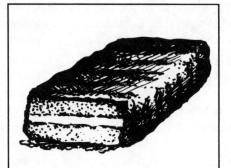

New-style crunchy milk chocolate biscuit with a delicious cream filling in 6 different flavours.

(a) Milk chocolate biscuit

CHARACTERISTICS OF THE MILK CHOCOLATE BISCUIT

- weighs 30g (approximately)
- approximate dimensions are 80mm x 30mm x 15mm
- crunchy biscuit
- covered in thick milk chocolate
- 6 different cream filling flavours: orange, coffee, cherry, peppermint, vanilla, chocolate

(b) Characteristics of the milk chocolate biscuit

INGREDIENTS OF THE MILK CHOCOLATE BISCUIT

- milk chocolate
- sugar
- wheat flour
- vegetable fat
- oatmeal
- butter-oil
- invert sugar syrup
- desiccated coconut
- whey powder
- skinned milk powder
- raising agents
- sodium bicarbonate
- ammonium bicarbonate
- salt
- molasses
- cream fillings
- flavouring

(c) Ingredients of the milk chocolate biscuit

NUTRITIONAL INFORMATION ABOUT THE MILK CHOCOLATE BISCUIT

Nutrition	Typical values	
	per 100g (3.5 oz)	per bar
Energy	500 kcal 2100 kJ	150 kcal 630 kJ
Protein	6.0g	1.8g
Carbohydrate of which	70.0g	21.0g
sugars	46.0g	13.8g
starch	24.0g	7.2g
Fat of which	22.0g	6.6g
saturates	13.0g	3.9g
mono-unsaturates	8.0g	2.4g
poly-unsaturates	1.0g	0.3g
Fibre	1.7g	0.5g
Sodium	0.3g	0.1g

(d) Table of nutritional information about the milk chocolate biscuit

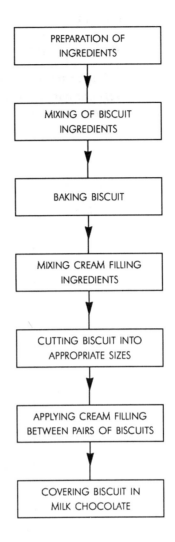

(e) A flow diagram of the manufacture
of the milk chocolate biscuit

There are many different ways that we can present the nutritional information. Before we start to look at a few types, let's produce the table of the nutritional information in a slightly different form.

NUTRITIONAL INFORMATION ABOUT THE MILK CHOCOLATE BISCUIT	
NUTRITIONAL FOOD COMPONENT	TYPICAL VALUES PER 100g
Protein	6.0g
Carbohydrate	70.0g
Fat	22.0g
Fibre	1.7g
Sodium	0.3g
TOTAL	100g

(f) Table of the nutritional information of 100g
of the milk chocolate biscuit

Let's produce the nutritional information in the form of a bar chart.

◆ Bar chart

A bar chart is a diagram which consists of a sequence of horizontal or vertical bars each representing the magnitude (size) of a quantity.

Let's produce a vertical bar chart for the nutritional information about the chocolate biscuit.

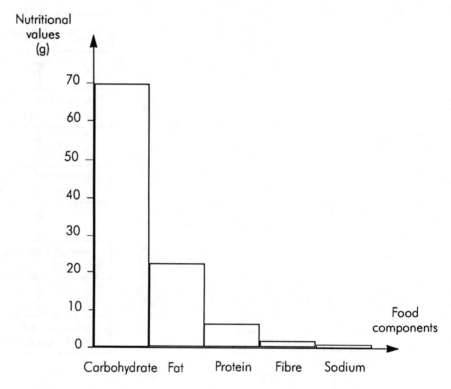

(g) Vertical bar chart of the distribution of nutrition in 100g of the chocolate biscuit

Right, now let's produce the nutritional information in the form of a pie chart.

◆ Pie chart

A pie chart is a circular diagram which represents a set of quantities as sectors. The area of each sector is proportional to the magnitude (size) of the quantity it is representing.

To produce a pie chart of the nutritional information, we need to calculate the following for each food component of the milk chocolate biscuit.

$$\text{Sector angle for food component} = \frac{\text{food component nutrition per 100g}}{100g} \times 360° \text{ (degrees in a circle)}$$

For example, let's calculate the sector angle for fat:

Sector angle for fat = $\dfrac{22.0}{100} \times 360° = 79.2°$

The complete table of sector angles is as follows

SECTOR ANGLES OF NUTRITIONAL INFORMATION	
NUTRITION	SECTOR ANGLE (DEGREES)
Protein	21.6°
Carbohydrate	252.0°
Fat	79.2°
Fibre	6.1°
Sodium	1.1°
TOTAL	360°

(h) Table of sector angles of the nutritional information

With the information in this form, we can now produce a pie chart.

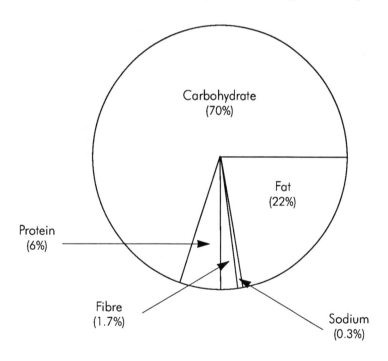

(i) Pie chart of the distribution of nutrition per 100g of the chocolate biscuit

Right, let's now look at the type of information that would be used in the presentation of a sleeveless cardigan design, as shown in (a) to (d).

➤ Sleeveless cardigan

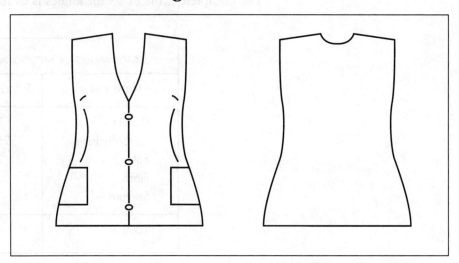

(a) The sleeveless cardigan, front and back

(b) Characteristics of the sleeveless cardigan

(c) Sleeveless cardigan pattern

CHARACTERISTICS OF THE
SLEEVELESS CARDIGAN

- semi-flared
- patch pockets

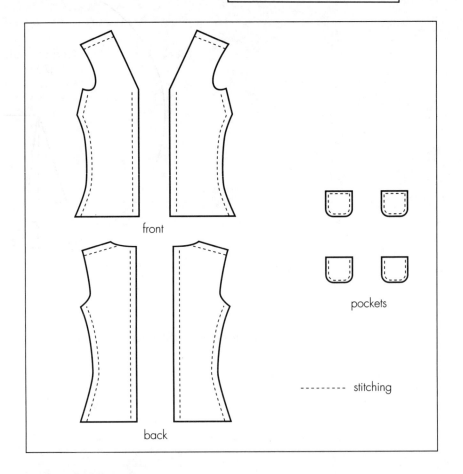

front

pockets

back

- - - - - - - stitching

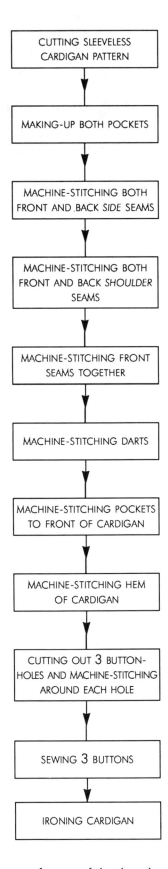

(d) Flow diagram of the manufacture of the sleeveless cardigan

And finally let's look at the type of information that would be used in the presentation of an energy-efficient immersion heater design, as shown in (a) to (e).

➤ Immersion heater

An immersion heater is an electrical device for heating the liquid in which it is immersed, usually under thermostatic control.

> ### CHARACTERISTICS
>
> - heating element is an alloy of aluminium, copper and molybdenum in the ratio of 4 : 1.5 : 4.5
> - heats a 60 litre tank to a temperature of 80°C in under 25 minutes
> - voltage: 240 volts AC
> - current drawn: 8.5 Amps
> - power consumption: 2.04 k Watts

(a) Characteristics of the heating element

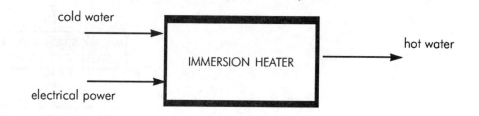

(b) Flow diagram of the immersion heater

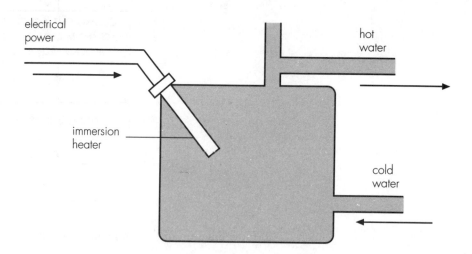

(c) Pictorial diagram of the immersion heater

	PROPOSED HEATING ELEMENT	TYPICAL HEATING ELEMENT
Time (min)	Temperature (°C)	Temperature (°C)
0	3	3
5	32	27
10	52	46
15	68	63
20	76	73
24	80	—
25		77
28		80

(d) Table comparing the times taken for the proposed heating element and a typical heating element to heat 60 litres of water to a temperature of

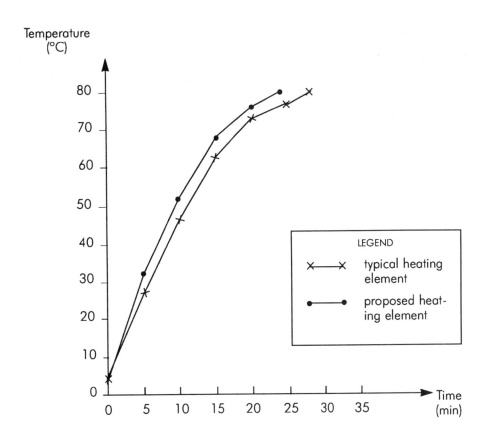

80°C

(e) Multiple line graph of the comparison in time of heating 60 litres of water up to 80°C using the proposed heating element and a typical heating element

This task is carried out in conjunction with tasks 27, 28, 29, 31, and 32

AUDIENCE

- familiar
- client

TASK 30

Present the product proposals to an audience

Use all the resources available to you.

Present one product proposal to an audience and the other to a client. The type of audience was decided upon in task 28.
Detailed and outline presentations were decided upon in task 27.

Comment briefly on how you felt your presentations went.

Keep a record of the comments in your portfolio.

FEEDBACK

- quantitative judgements
- qualitative judgements

EXAMPLES

EXAMPLES

2.11 FEEDBACK

At the end of the presentation, both quantitative judgements and qualitative judgements are fed back from the audience to the presenter (designer).

➤ Quantitative judgements

These involve or relate to considerations of amount or size.

For example, it may be judged that: there is too large a fat content in a food product, or that the length of the stool legs could be 2 or 3 cm longer.

➤ Qualitative judgements

These involve or relate to distinctions based on quality or qualities.

For example, it may be judged that: the finish of the jacket is not of a sufficiently high standard or that the screw holes on the bookcase need plugging.

This task is carried out in conjunction with tasks 27, 28, 29, 30, and 32

FEEDBACK
- quantitative judgements
- qualitative judgements

TASK 31
Obtaining sufficient feedback to revise proposals

Use all the resources available to you, particularly task 30.

Hand out the feedback form, devised in task 27, prior to the start of both of your presentations and ask the audience to complete it at the end of your presentations.

Remember to ask the audience for both quantitative and qualitative judgements.

Keep a record of their judgements in your portfolio.

This task is carried out in conjunction with tasks 27, 28, 29, 30, and 31

TASK 32

Complete the final product proposal

Use all the resources available to you, particularly task 31.

Study the feedback provided by each audience and where relevant modify the product proposal.

Produce a completed product proposal for each of your two products.

Show how you modified each product proposal.

Bring together tasks 27–32 inclusive to produce a report titled: 'Finalise proposals using feedback from presentations'.

Keep the report in your portfolio.

3 · PRODUCTION PLANNING, COSTING AND QUALITY ASSURANCE

3.1 PRODUCING A PRODUCTION PLAN

PRODUCT SPECIFICATION

- product description
- materials
- measurements
- critical control points
- finish
- quality indicators

KEY STAGES OF PRODUCTION

- material preparation
- processing
- assembling
- finishing
- packaging

RESOURCES

- capital
- human
- tooling-up
- services

PRODUCTION SCHEDULE

- sequence of jobs
- processing times
- critical control points

Producing a production plan

What is a production plan? A production plan for the manufacture of a product includes the following:

- a plan of what production facilities are needed;
- a plan of the siting of these facilities;
- a plan of the layout of these facilities;
- a plan of how the facilities are used in the manufacture of a product.

How do you produce a production plan?

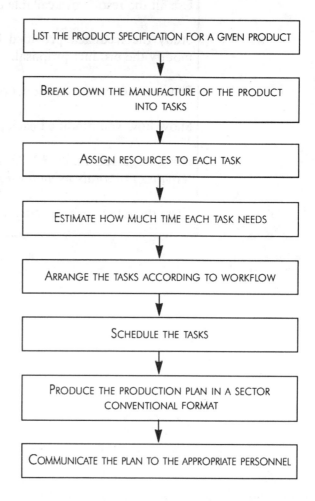

LIST THE PRODUCT SPECIFICATION FOR A GIVEN PRODUCT

↓

BREAK DOWN THE MANUFACTURE OF THE PRODUCT INTO TASKS

↓

ASSIGN RESOURCES TO EACH TASK

↓

ESTIMATE HOW MUCH TIME EACH TASK NEEDS

↓

ARRANGE THE TASKS ACCORDING TO WORKFLOW

↓

SCHEDULE THE TASKS

↓

PRODUCE THE PRODUCTION PLAN IN A SECTOR CONVENTIONAL FORMAT

↓

COMMUNICATE THE PLAN TO THE APPROPRIATE PERSONNEL

3.2 PRODUCT SPECIFICATIONS

DEFINITION
Specification: a detailed description of the criteria for the materials, construction, finish, and performance of a product.

The first step in producing a production plan is to list the details of the product specification.

What is a product specification?

There are a number of different types of specification, of which the most important is the product specification.

A product specification is a document which gives a detailed description of the product, including dimensions, materials, critical control points, finish and quality indicators.

The general layout of any type of specification is as follows:

- Title
- Contents
- Introduction
- Role of product
- Related documentation
- Technical specification
- Information
- Appendices
- Supplementary information

A product specification could contain any of the following information:

– design	– maintenance
– construction	– performance
– materials	– spares
– manufacture	– power ratings
– reliability	– measurements
– conformance to relevant safety standards	– weight
– conformance to national and international standards and specifications	– finish
	– life expectancy
	– quality assurance
– acceptance conditions	– packaging
– despatch	– installation
– drawings	– ordering

PRODUCT SPECIFICATION

- product description
- materials
- measurements
- critical control points
- finish
- quality indicators

Let's take a look at a simple example of the product specification for a galvanized steel metal tray.

➤ Product specification for a metal tray

◆ Product description

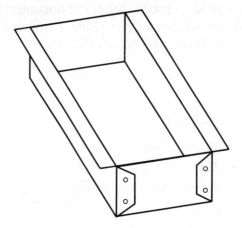

The product is a rectangular galvanized steel metal tray, 500mm x 360mm, and is 50mm deep with a 20mm fold along its length.

◆ Materials

Sheet metal, 500mm × 500mm per tray.

◆ Measurements

Tolerances ± 1mm.

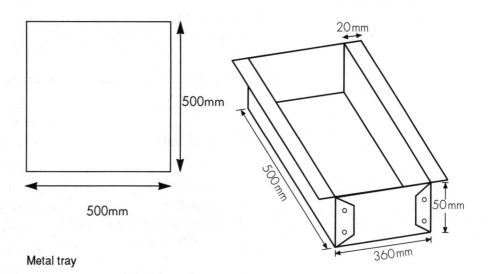

Metal tray

◆ Critical control points

During the manufacture of a product there will be a number of points, known as critical control points, where the product must be checked to make sure that it will meet its specification.

There are three critical control points in the manufacture of the metal tray:

1 The quality of the sheet metal is checked upon collection from the material stores prior to the start of the manufacturing process.

2 The dimensions of the tray are checked when it has been folded into shape.

3 The quality of the finished product is checked after it has been galvanized.

◆ Finish

The metal tray is finished off with a protective zinc coating, known as galvanization.

◆ Quality indicators

We first met quality indicators in section 1.7 and found that they were a variable or attribute of a product that could be measured or assessed respectively.

The data obtained is compared against the product's specification to give an indication of its quality.

At the three critical control points, the following types of quality indicator will be applied:

CRITICAL CONTROL POINT	QUALITY INDICATOR
1	Appearance
2	Size
3	Appearance/functionality

This task is carried out in conjunction with tasks 34, 35, 36, 37, and 38

SPECIFICATION DETAILS

- product description
- materials
- measurements
- critical control points
- finish
- quality indicators

TASK 33

Listing the specification details of a product

To carry out this task, use all the resources available to you – such as the library, magazines, journals, teachers, tutors, local companies and 'yellow pages'.

By task 32 you had produced completed product proposals for your two selected products.

Identify and list the specification details for both your selected products.

Keep the specification details in your portfolio. These will form part of the production plans which will be put together in task 38.

KEY PRODUCTION STAGES

• material preparation
• processing
• assembly
• finishing
• packaging

3.3 KEY PRODUCTION STAGES

We met the key production stages of manufacturing a product in section 1.4. You remember the production stages required to manufacture a metal tray:

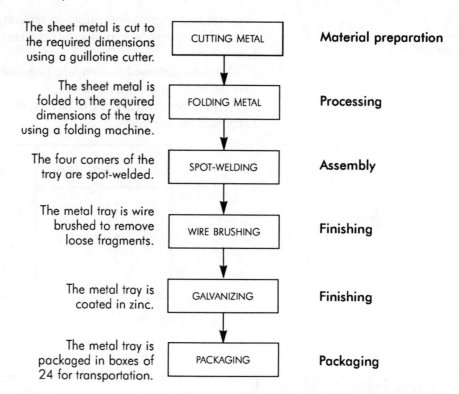

The sheet metal is cut to the required dimensions using a guillotine cutter.	CUTTING METAL	**Material preparation**
The sheet metal is folded to the required dimensions of the tray using a folding machine.	FOLDING METAL	**Processing**
The four corners of the tray are spot-welded.	SPOT-WELDING	**Assembly**
The metal tray is wire brushed to remove loose fragments.	WIRE BRUSHING	**Finishing**
The metal tray is coated in zinc.	GALVANIZING	**Finishing**
The metal tray is packaged in boxes of 24 for transportation.	PACKAGING	**Packaging**

Key production stages in the manufacture of a metal tray

This task is carried out in conjunction with tasks 33, 35, 36, 37, and 38

KEY PRODUCTION STAGES

• material preparation
• processing
• assembling
• finishing
• packaging

TASK 34

Describing the key stages of production in the manufacture of a product

To carry out this task, use all the resources available to you.

Describe and produce a flow diagram of the key production stages required to manufacture each of your two selected products.

Keep the description and flow diagrams of the key stages of production in your portfolio.

These will form part of the production plans which will be put together in task 38.

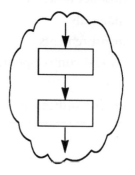

Flow diagram

3.4 RESOURCE REQUIREMENTS

To manufacture a product many different types of resource are required – such as the materials that make up the product and the machinery that processes the materials into a product.

Let's begin by looking at capital resources.

➤ Capital resources

Capital resources involved in the production process are:

- plant;
- equipment;
- machinery.

➤ Human resources

Human resources involved, both directly and indirectly, in the production process are:

- direct production workers;
- technical support;
- management.

➤ Material resources

Material resources involved in the production process are:

- raw materials, including bought-in components;
- consumables.

➤ Tooling-up resources

Tooling-up resources involved in the production process are, for example:

- drilling tools used in drilling machines;
- needles used in sewing machines;
- cutting tools used in lathes and milling machines.

➤ Services

Services resources involved in the production process are, for example:

- gas;
- water;
- electricity;
- compressed air.

RESOURCES

- capital
- human
- material
- tooling-up
- services

DEFINITION
Plant: production facility.

DEFINITION
Equipment: tools and apparatus.

DEFINITION
Machinery: machines used in the production process.

Let's determine the resource requirements for the manufacture of 200 galvanized steel metal trays.

TYPE OF PRODUC-TION STAGE	RESOURCE TYPE / PRO-DUCTION STAGE	CAPITAL RESOURCES	HUMAN RESOURCES	MATERIAL RESOURCES	TOOLING RESOURCES	SERVICES RESOURCES
MATERIAL PREPARATION	Cutting metal to shape	2 guillotine cutters	2 production workers	Sheet metal	—	Electricity
PROCESSING	Folding metal to shape	2 folding machines	2 production workers	—	—	Electricity
ASSEMBLY	Spot-welding corner joint flaps	2 spot-welding equipments	2 production workers	Weld tips	—	Electricity
FINISHING	Wire brushing metal tray	1 wire brush equipment	1 production worker	—	Wire brush	Electricity
FINISHING	Galvanizing metal tray	—	1 production worker	—	—	Electricity
PACKAGING	Packaging metal tray	Scissors	1 production worker	Packing paper, boxes, tape	—	—

Resource requirements for the manufacture of 200 metal trays

This task is carried out in conjunction with tasks 33, 34, 36, 37, and 38

RESOURCE REQUIREMENTS

- capital
- human
- material
- tooling-up
- services

TASK 35

Identifying the resource requirements to manufacture a product

Use all the resources available to you.

Identify and produce a table of the resource requirements to manufacture each of your two selected products.

Discuss and agree with your teacher/tutor a sensible product output figure, ie the number of items to manufacture in a given time for each of your two selected products. Perhaps an output of between 50 and 500 products in 1–4 weeks could be agreed.

Identify and produce a table of the resource requirements needed to manufacture each of your two selected products to the agreed product outputs.

Keep the tables of resource requirements in your portfolio.

These will form part of the production plans which will be put together in task 38.

3.5 PROCESSING TIMES

The production stage processing time can be calculated using the following formula:

$$\begin{array}{l}\text{Production} \\ \text{stage} \\ \text{timing} \end{array} = \begin{array}{l}\text{Machine} \\ \text{setting - up} \\ \text{time} \end{array} + \left(\begin{array}{l}\text{Operation time} \\ \text{per item} \end{array} \times \begin{array}{l}\text{Number of} \\ \text{items processed} \end{array} \right)$$

Please note:

- all timings must be in the *same* units, normally minutes or seconds;
- machine setting-up time normally only occurs once per production run;
- if identical processes are operating concurrently (ie taking place at the same time), then we assume that they take the same amount of time to set up and to carry out each operation.

DEFINITION
Concurrent: taking place at the same time.

Let's look at some examples:

EXAMPLE

An operator on a sewing machine takes three minutes to set up the machine and 48 seconds to carry out each operation.
   What is the production stage time for one operation and for 500 operations?

For one operation:

$$\begin{array}{c} \text{Production} \\ \text{stage} \\ \text{timing} \end{array} = \begin{array}{c} \text{Machine} \\ \text{setting-up} \\ \text{time} \end{array} + \left(\begin{array}{c} \text{Operation time} \\ \text{per item} \end{array} \times \begin{array}{c} \text{Number of} \\ \text{items processed} \end{array} \right)$$

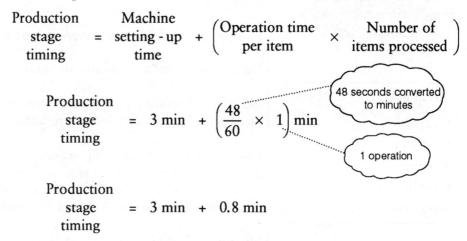

$$\begin{array}{c} \text{Production} \\ \text{stage} \\ \text{timing} \end{array} = 3 \text{ min} + \left(\frac{48}{60} \times 1 \right) \text{min}$$

48 seconds converted to minutes

1 operation

$$\begin{array}{c} \text{Production} \\ \text{stage} \\ \text{timing} \end{array} = 3 \text{ min} + 0.8 \text{ min}$$

Production stage timing for 1 operation = 3.8 minutes.

For 500 operations:

$$\begin{array}{c} \text{Production} \\ \text{stage} \\ \text{timing} \end{array} = \begin{array}{c} \text{Machine} \\ \text{setting-up} \\ \text{time} \end{array} + \left(\begin{array}{c} \text{Operation time} \\ \text{per item} \end{array} \times \begin{array}{c} \text{Number of} \\ \text{items processed} \end{array} \right)$$

48 seconds converted to minutes

$$\begin{array}{c} \text{Production} \\ \text{stage} \\ \text{timing} \end{array} = 3 \text{ min} + \left(\frac{48}{60} \times 500 \right) \text{min}$$

500 operations

$$\begin{array}{c} \text{Production} \\ \text{stage} \\ \text{timing} \end{array} = 3 \text{ min} + 400 \text{ min}$$

Production stage timing for 500 operations = 403 minutes (6 hr 43 min).

Please note that:

$$\begin{array}{c} \text{Production} \\ \text{stage} \\ \text{timing} \end{array} \text{ for 500 operations} \neq \begin{array}{c} \text{Production} \\ \text{stage} \\ \text{timing} \end{array} \text{ for one operation} \times 500$$

because the machine only has to be set up once in both cases.

Now let's look at some concurrent processes.

If we have *two* identical processes operating at the same time, then each process will only have to carry out *half* the total operations required.

If we have *three* identical processes operating at the same time, then each process will only have to carry out *one third* of the total operations required.

And so on.

Let's look at another example.

An operator on a process for manufacturing plastic cups takes 2½ minutes to set up the machinery and 30 seconds to process one plastic cup.

What is the production stage timing for one operation and for 800 operations?

Go back to the formula:

$$\text{Production stage timing} = \text{Machine setting-up time} + \left(\text{Operation time per item} \times \text{Number of items processed}\right)$$

For 1 operation

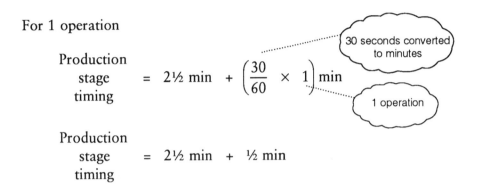

$$\text{Production stage timing} = 2½ \text{ min} + \left(\frac{30}{60} \times 1\right) \text{ min}$$

$$\text{Production stage timing} = 2½ \text{ min} + ½ \text{ min}$$

Production stage timing for 1 operation = 3 minutes.

For 800 operations, using the same formula:

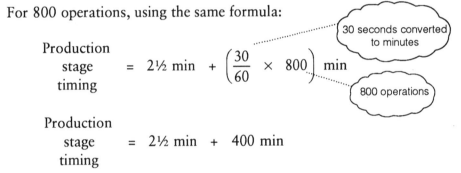

$$\text{Production stage timing} = 2½ \text{ min} + \left(\frac{30}{60} \times 800\right) \text{ min}$$

$$\text{Production stage timing} = 2½ \text{ min} + 400 \text{ min}$$

Production stage timing for 800 operations = 402.5 minutes (6 hr 42½ min).

Now suppose we have two processes operating at the same time: then how long will it take to manufacture 800 plastic cups?

Each process will manufacture 400 plastic cups.
Using the formula again:

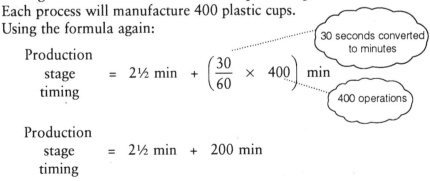

$$\text{Production stage timing} = 2½ \text{ min} + \left(\frac{30}{60} \times 400\right) \text{ min}$$

$$\text{Production stage timing} = 2½ \text{ min} + 200 \text{ min}$$

Production stage timing for 800 operations using two processes = 202.5 minutes (3 hr 22½ min).

We are now ready to calculate the production stage timings to manufacture the 200 steel trays.

Let's look at an example.

Two production workers each use a guillotine cutter to cut the metal to shape. If it takes 5 minutes to set up a cutting machine and 2½ minutes to carry out one operation, then how long will it take to process 200 items?

Since we have two processes, then each process only has to carry out 100 operations.

$$\text{Production stage timing} = \text{Machine setting-up time} + \left(\text{Operation time per item} \times \text{Number of items processed}\right)$$

$$\text{Production stage timing} = 5 \text{ min} + (2\tfrac{1}{2} \times 100) \text{ min}$$

$$\text{Production stage timing} = 5 \text{ min} + 250 \text{ min}$$

Therefore:
Production stage timing for 200 operations using 2 cutters = 255 minutes (4¼ hr).

Let's look at the production stage timings to manufacture 200 metal trays.

Remember the formula:

$$\text{Production stage timing} = \text{Machine setting-up time} + \left(\text{Operation time per item} \times \text{Number of items processed}\right)$$

On the opposite page is a chart showing the production stage timing to manufacture 200 metal trays

TYPE OF PRODUCTION STAGE	PRODUC-TION STAGE	MACHINE SET-UP TIME (min) (A)	NUMBER OF OPERATIONS (B)	OPERATION TIME PER TRAY (min) (C)	TOTAL OPERATION TIME (min)	PRODUCTION STAGE TIME (min) (A) + (B x C)
MATERIAL PREPARATION	Cutting metal ①	5	100	2.5	100 x 2.5 = 250	5 + 250 = 255
	Cutting metal ②	5	100	2.5	100 x 2.5 = 250	
PROCESSING	Folding metal ①	5	100	3.5	100 x 3.5 = 350	5 + 350 = 355
	Folding metal ②	5	100	3.5	100 x 3.5 = 350	
ASSEMBLY	Spot-welding ①	10	100	0.9	100 x 0.9 = 90	10 + 90 = 100
	Spot-welding ②	10	100	0.9	100 x 0.9 = 90	
FINISHING	Wire brushing	5	200	1.4	200 x 1.4 = 280	5 + 280 = 285
FINISHING	Galvanizing	10	200	0.8	200 x 0.8 = 160	10 + 160 = 170
PACKAGING	Packaging	5	200	0.5	200 x 0.5 = 100	5 + 100 = 105
TOTAL PRODUCTION TIME						1270 minutes (21 hr 10 min)

Production stage timings to manufacture 200 trays

This task is carried out in conjunction with tasks 33, 34, 35, 37, and 38

TASK 36

Estimating the processing times for each key stage of production

Use all the resources available to you.

Estimate the processing times for each key stage of production needed to manufacture each of your two selected products.

You will need to estimate the setting-up time and process time per product in order to be able to find the overall process time per product at each key stage of production.

In task 35, you produced tables of resource requirements needed to manufacture each of your two selected products to their agreed product outputs. Use these tables to estimate the processing times for each key stage of production required to achieve this manufacturing target.

Produce a table of the estimated processing times for each key stage of production required to manufacture your two selected products to their agreed product outputs.

Keep your tables of estimated processing times in your portfolio.

These will form part of the production plans which will be put together in task 38.

FACTORS TO BE CONSIDERED WHEN SCHEDULING

- labour availability
- machine availability
- delivery date
- customer
- costs

3.6   PRODUCTION SCHEDULES

A production schedule is the sequencing of operations in the most cost-effective way within a factory.

A very common method of preparing a production schedule is using a Gantt chart.

➤ Gantt chart

A Gantt chart, as shown on the next page, represents the activities required to manufacture a product in a diagrammatic form against a time-scale. It is an ideal method of displaying and communicating production schedules: easy to read, and easy to change and update.

Let's look at a Gantt chart for scheduling the manufacture of the 200 metal trays.

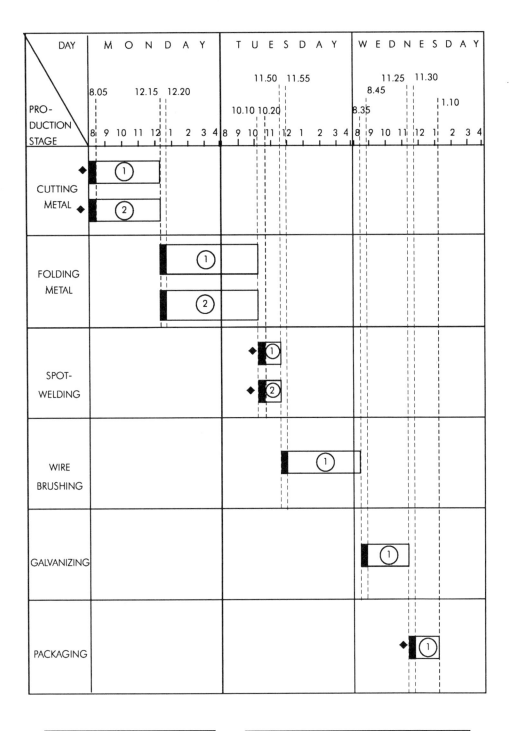

Production schedule for manufacturing 200 metal trays
Gantt chart

ASSUMPTIONS

(i) Setting-up time occurs at the start of each process

(ii) Zero transit time between processes

Set-up time   ① Worker 1

Process time   ② Worker 2

◆ Critical control points

In section 3.5, the production stage timings for the manufacture of 200 metal trays were calculated. Now let's see how we used this information to produce the Gantt chart on the previous page.

The first production stage is material preparation, which involves cutting the metal to the required shape.

TYPE OF PRODUCTION STAGE	PRODUC-TION STAGE	MACHINE SET-UP TIME (min) (A)	NUMBER OF OPERATIONS (B)	OPERATION TIME PER TRAY (min) (C)	TOTAL OPERATION TIME (min) (B x C)	PRODUCTION STAGE TIME (min) (A) + (B x C)
Material preparation	Cutting metal ①	5	100	2.5	100 x 2.5 = 250	5 + 250 = 255
	Cutting metal ②	5	100	2.5	100 x 2.5 = 250	5 + 250 = 255

(We assume that the factory operates continuously from 8 am to 4 pm Monday to Friday.)

Cutting begins at 8 am on Monday morning and two cutters operate at the same time.
  Set-up time for both cutters takes 5 minutes from 8 am to 8.05 am (We assume that the setting-up of the machinery occurs at the start of each process.)
  Each cutter carries out half the total operations, ie 100 operations taking 250 minutes.

$$250 \text{ minutes} = \frac{250}{60} \text{ hours} = 4 \text{ hours } 10 \text{ mins}$$

Therefore each cutter operates from 8.05 am to 12.15 pm.

The next production stage is folding the metal and this will be scheduled to start at 12.15 pm. (We assume that there is zero transit time between each production stage.)
  Set-up time for both folders takes 5 minutes, and will take from 12.15 pm to 12.20 pm. Each folder carries out half the total operations, ie 100 operations taking 350 minutes.

$$350 \text{ minutes} = \frac{350}{60} \text{ hours} = 5 \text{ hours } 50 \text{ mins}$$

The folders operate from 12.20 pm until the factory closes at 4 pm on the Monday, which is 3 hours 40 minutes.
  The folders will need to operate for another 2 hours 10 minutes on Tuesday morning to complete the folding operations.
(3 hours 40 minutes + 2 hours 10 minutes = 5 hours 50 minutes)
  Folding continues from 8 am to 10.10 am on Tuesday morning.

Each production stage is added in order to the Gantt chart until finally at 1.10 pm on Wednesday afternoon the 200 metal trays will be manufactured. We now have a production schedule for manufacturing the 200 metal trays.

Production schedules can be prepared by computer or manually. The manual preparation of schedules is becoming rare and is only used when the manufacture of a product requires few operations.

Scheduling is ideally suited for computers and Gantt chart software packages are readily available. These packages allow you to output data according to your needs, such as:

- print all view screens and charts;
- plot Gantt charts;
- create customized reports by specifying which data should be printed and how it should be laid out.

This task is carried out in conjunction with tasks 33, 34, 35, 36, and 38

PRODUCTION SCHEDULE

- sequence of jobs
- processing times
- critical control points

TASK 37

Producing a production schedule

Use all the resources available to you, particularly tasks 36 and 33.

Produce production schedules for the manufacture of both your selected products to their agreed product outputs. Use the tables of processing times that you produced in task 36.

Critical control points were identified in task 33.

Keep the production schedules in your portfolio.

These will form part of the production plans which will be put together in task 38.

3.7 PRODUCING THE PRODUCTION PLAN

A production plan for a product includes:

- identification of the product specification details including critical control points;
- description of the key production stages;
- identification of the resources required;
- estimation of processing times for each key production stage;
- production schedule.

Let's look at the production plan for the manufacture of 800 cotton sports skirts.

➤ Product specification details

◆ Product description

Plain grey cotton sports skirt in 4 sizes: Small, Medium, Large, and Extra large.

◆ Materials

Grey cotton roll, zip, and a button.

◆ Measurements

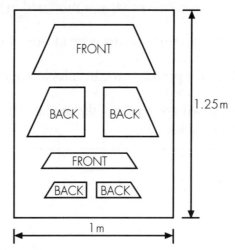

Skirt pattern

◆ Critical control points

1 The quality of the cotton material is checked after it is cut into pre-determined lengths.

2 The quality of the finished product is checked prior to packaging.

◆ Finish

The cotton sports skirt will be labelled and ironed.

◆ Quality indicators

CRITICAL CONTROL POINT	QUALITY INDICATOR
1	Appearance
	Touch
2	Appearance
	Functionality

Functionality at critical control point 2 is the testing of the zipper

➤ Key production stages

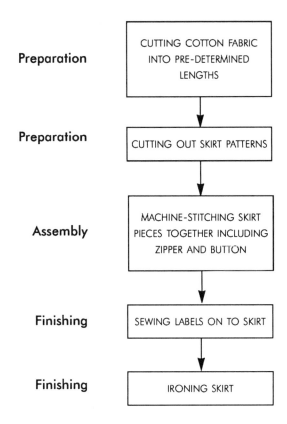

Preparation	CUTTING COTTON FABRIC INTO PRE-DETERMINED LENGTHS
Preparation	CUTTING OUT SKIRT PATTERNS
Assembly	MACHINE-STITCHING SKIRT PIECES TOGETHER INCLUDING ZIPPER AND BUTTON
Finishing	SEWING LABELS ON TO SKIRT
Finishing	IRONING SKIRT

Key production stages in the manufacture of a sports skirt

◆ Resource requirements

TYPE OF PRODUC-TION STAGE / RESOURCE TYPE PRO-DUCTION STAGE		CAPITAL RESOURCES	HUMAN RESOURCES	MATERIAL RESOURCES	TOOLING RESOURCES	SERVICES RESOURCES
MATERIAL PREPARATION	Cutting cotton fabric to pre-deter-mined lengths	Scissors	2 operatives	Cotton roll	—	—
PROCESSING	Cutting out skirt patterns for each of 4 sizes	Pattern-cutting machine	1 operative	—	—	Electricity
ASSEMBLY	Machine-stitching skirt pieces together	Sewing machine	4 operatives	Sewing cotton	Machine needle	Electricity
FINISHING	Labelling skirt	Sewing machine	4 operatives	Labels	—	Electricity
FINISHING	Ironing skirt	Iron	2 operatives	—	—	Electricity
PACKAGING	Packaging skirt	Scissors	2 operatives	Packing paper, boxes, tape	—	—

◆ Processing times

TYPE OF PRODUCTION STAGE	PRODUC- TION STAGE	MACHINE SET-UP TIME (min) (A)	NUMBER OF OPERATIONS (B)	OPERATION TIME PER SKIRT (min) (C)	TOTAL OPERATION TIME (min)	PRODUCTION STAGE TIME (min) (A) + (B x C)
MATERIAL PREPARATION	Cutting cotton fabric to pre-deter- mined lengths		400 400	0.5 0.5	400 x 0.5 = 200 400 x 0.5 = 200	200
PROCESSING	Cutting out skirt pattern for each of 4 sizes	30 (4 different sizes)	8 (100 skirts at a time)	10 (100 skirts at a time)	8 x 10 = 80	30 + 80 = 110
ASSEMBLY	Machine- stitching skirt pieces together	5 5 5 5	200 200 200 200	5 5 5 5	200 x 5 = 1000 200 x 5 = 1000 200 x 5 = 1000 200 x 5 = 1000	5 +1000 = 1005
FINISHING	Labelling skirt	5 5 5 5	200 200 200 200	0.5 0.5 0.5 0.5	200 x 0.5 = 100 200 x 0.5 = 100 200 x 0.5 = 100 200 x 0.5 = 100	5 +100 = 105
FINISHING	Ironing skirt	5 5	400 400	1 1	400 x 1 = 400 400 x 1 = 400	400 +5 = 405
PACKAGING	Packaging skirt		400 400	0.2 0.2	400 x 0.2 = 80 400 x 0.2 = 80	80
TOTAL PRODUCTION TIME						1905 minutes (31 hr 45 min)

Processing times for manufacturing 800 sports skirts

Production schedule for manufacturing 800 sports skirts

ASSUMPTIONS

(i) Setting-up time occurs at the start of each-process

(ii) Zero transit time between processes

■ Set-up time   ① Worker 1

▯ Process time   ② Worker 2

               ③ Worker 3

               ④ Worker 4

◆ Critical control points

This task is carried out in conjunction with tasks 33, 34, 35, 36, and 37

TASK 38

Producing a production plan

Use all the resources available to you, particularly tasks 33–37 inclusive.

Produce production plans for the manufacture of both your selected products to their agreed product outputs.

They should include the following:

- identification of the product specification details from task 33;
- description and flow diagram of the key production stages from task 34;
- identification of the resource requirements from task 35;
- estimation of the processing times for each key production stage from task 36;
- the production schedule from task 37.

Keep the completed production plans in your portfolio.

DIRECT COSTS

- human resources
- material resources
- services resources

3.8 DIRECT COSTS

The direct costs for a stage of production are those costs which can be specifically identified with that stage of production.

The direct costs for each production stage can be found using the formula:

$$\text{Production stage direct costs} = \text{Human resource costs} + \text{Material resource costs} + \text{Services resource costs}$$

Let's begin by looking at the human resource costs.

➤ Human resource costs

These are the cost of the wages paid to the personnel working directly on the production stage.

To calculate the cost of each human resource we can use the following formula:

$$\text{Cost of human resource} = \text{Time spent producing in hours} \times \text{Hourly rate for producing}$$

Let's look at some examples of calculating human resource costs.

EXAMPLE

Suppose a sewing machine operator takes 403 minutes to carry out 500 operations. If the operator is paid £3.50 per hour, then what are the human resource costs?

$$\text{Cost of human resource} = \text{Time spent producing in hours} \times \text{Hourly rate for producing}$$

$$\text{Cost of human resource} = \frac{403}{60} \times £3.50$$

403 minutes converted to hours

Cost of human resource = £23.51.

EXAMPLE

Suppose two operatives mix cake ingredients for 3 hours 25 minutes and are both paid £3.68 per hour, then what are the human resource costs?

Let's calculate the costs for one operative and then double the answer.

$$\text{Cost of human resource} = \text{Time spent producing in hours} \times \text{Hourly rate for producing}$$

$$\text{Cost of human resource} = 3\frac{25}{60} \times £3.68$$

25 minutes converted to hours

Cost of human resource = £12.57.

So for both operatives:

Cost of human resources = £12.57 x 2 = £25.14.

Now let's consider the material resource costs.

➤ Material resource costs

Material resource costs are those costs attributed to the raw materials, components, consumables, and packaging used at a production stage.

To calculate the cost of each material resource, we can use the following formula:

$$\text{Cost of material resources} = \text{Material cost of one item} \times \text{Number of items produced}$$

Let's look at some examples of calculating material resource costs.

EXAMPLE

Suppose 1 000 plastic cups are being produced by a process of moulding, and the raw materials required to produce one cup costs 0.15p, then what are the material resource costs?

$$\text{Cost of material resources} = \text{Material cost of one item} \times \text{Number of items produced}$$

$$\text{Cost of material resources} = 0.15p \times 1\ 000$$

Cost of material resources = £1.50.

Here is another example.

EXAMPLE

At a particular production stage in a local joinery, 2.8 metres of 25.4mm x 25.4mm wood is used in the manufacture of a stool. If the cost of the wood is 46p per metre, then calculate the cost of material resources for one stool, and also for 40 stools at that production stage.

Let's determine the cost of material resources for one stool.

$$\text{Cost of material resources for one stool at a particular production stage} = \text{Length of 25.4mm x 25.4mm wood used} \times \text{Cost of 25.4mm x 25.4mm wood used}$$

$$\text{Cost of material resources per stool} = 2.8 \text{ metres} \times 46p \text{ per metre}$$

Cost of material resources = £1.29.

And hence the cost of material resources for 40 stools at that particular production stage is:

$$\text{Cost of material resources} = \text{Material cost of one stool} \times \text{Number of stools produced}$$

$$\text{Cost of material resources} = £1.29 \times 40$$

Cost of material resources = £51.60.

And as a last example, let's look at manufacturing a galvanized sheet metal tray.

At the first production stage the metal is cut to the required dimensions of 0.5 metres by 0.5 metres.

Let's begin by calculating the cost of the metal for one tray.

The sheet metal is collected from the stores in 1 metre x 1 metre strips. The cost of the sheet metal is £4.32 per metre squared (m²)
How many trays can we make from 1 m² of metal?

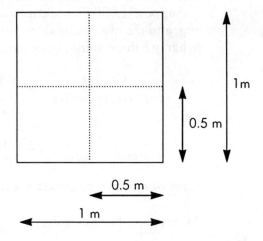

$$\text{Number of trays} = \frac{\text{Area of sheet metal}}{\text{Area of metal tray}}$$

$$\text{Number of trays} = \frac{1m \times 1m}{0.5m \times 0.5m} = \frac{1m^2}{0.25m^2}$$

Number of trays = 4

The cost of sheet metal is £4.32 per m²

$$\text{The cost of sheet metal per tray} = \frac{£4.32}{4} = £1.08$$

Therefore cost of material resources = £1.08.

And finally let's consider services resource costs.

➤ Services resource costs

Services resource costs are those costs attributed to the electricity, water, and gas that may be used at a production stage.

To calculate the cost of services resources we can use the following formula:

$$\text{Cost of services resources} = \text{Units consumed} \times \text{Cost per unit}$$

Let's look at some examples of calculating service resource costs.

EXAMPLE

A production stage in a toy factory consumes 0.85 units of electricity in the manufacture of a spinning top. If the cost of one unit of electricity is 7.17p, then calculate the cost of the electricity used per product at that particular production stage.

$$\text{Cost of electricity consumed} = \text{Number of units consumed} \times \text{Cost per unit}$$

$$\text{Cost of electricity consumed} = 0.85 \times 7.17\,\text{p}$$

Cost of electricity consumed = 6.1p.

The unit of electricity is the kWh which stands for kilowatt hour. The kWh is the rate of working of 1 000 watts for one hour.
   This means that in the previous example a consumption of 0.85 units is equivalent to 0.85 kWh or a rate of working of 850 watts for one hour.

Let's look at another example.

EXAMPLE

An item of electrical equipment used at a production stage consumes 2 kilowatts of electrical power. If the electrical equipment takes 6 minutes to process each product and the cost of one unit of electricity is 7.17p, then calculate the cost of electricity used per product.

Firstly, we need to calculate the number of units consumed per product.
   The manufacture of 1 product consumes 2 kilowatts for 6 minutes.

Consumption of 2 kW for 6 minutes

$$= \text{Consumption of 2kW for } \frac{6}{60} \text{ hours}$$

$$= \text{Consumption of 2kW for 0.1 hours}$$

$$= 0.2 \text{ kWh}$$

Now 1 kWh = 1 unit. Therefore 0.2 kWh = 0.2 units. So 0.2 units of electricity are consumed in manufacturing each product.

$$\begin{matrix} \text{Cost of} \\ \text{electricity} \\ \text{consumed} \end{matrix} = \begin{matrix} \text{Number of} \\ \text{units} \\ \text{consumed} \end{matrix} \times \begin{matrix} \text{Cost} \\ \text{per unit} \end{matrix}$$

$$\begin{matrix} \text{Cost of} \\ \text{electricity} \\ \text{consumed} \end{matrix} = 0.2 \times 7.17\text{p}$$

Cost of electricity consumed = 1.43p.

And here is another example, this time looking at gas consumption.

EXAMPLE

A production stage in an engineering company uses 2.74 kWh of gas in the manufacture of a metal product. If the cost of 1kWh of gas is 1.477p, then calculate the cost of gas used per product at that particular production stage.

$$\begin{matrix} \text{Cost of} \\ \text{gas consumed} \end{matrix} = \begin{matrix} \text{Number of} \\ \text{kWh consumed} \end{matrix} \times \begin{matrix} \text{Cost} \\ \text{per kWh} \end{matrix}$$

$$\begin{matrix} \text{Cost of gas} \\ \text{consumed} \end{matrix} = 2.74 \times 1.477\text{p}$$

Cost of gas consumed = 4.05p.

And a final example:

EXAMPLE

A production stage in a food processing factory uses 6.8m³ of water to rinse the equivalent of 300 tins of processed food. If the cost for each cubic metre of water is 71.43p, then calculate the cost of water used to rinse the equivalent of 1 tin of processed food.

The rinsing of the equivalent of 1 tin of processed food takes:

DEFINITION
m³: a cubic metre

$$\frac{6.8\text{m}^3}{300} = 0.02267 \ \text{m}^3 \text{ of water}$$

Therefore:

$$\begin{matrix} \text{Cost of water} \\ \text{consumed} \end{matrix} = \begin{matrix} \text{m}^3 \text{ of water} \\ \text{consumed} \end{matrix} \times \begin{matrix} \text{Cost per m}^3 \\ \text{of water} \end{matrix}$$

$$\begin{matrix} \text{Cost of water} \\ \text{consumed} \end{matrix} = 0.02267 \ \text{m}^3 \times 71.43\text{p per m}^3$$

Cost of water consumed = 1.62p.

Let's now calculate the direct costs for the manufacture of the 800 skirts that we introduced earlier in the chapter.

On the next page you will see a chart using the processing times that we calculated for each production stage in section 3.7 used to calculate the direct costs.

TYPE OF PRODUCTION STAGE	PRODUCTION STAGE	HUMAN RESOURCE COSTS (A)	MATERIAL RESOURCE COSTS (B)	SERVICES RESOURCE COSTS (C)	PRODUCTION STAGE COSTS (A + B + C)
MATERIAL PREPARATION	Cutting cotton fabric to pre-determined lengths	2 operatives @ £3.80 per hour for 200 min Cost = (200÷60) x £3.80 x 2 Cost = £25.33 (2 ops)	800 skirts @ 1.25m x 1m = 1000m^2 Cotton fabric is £2.25 per m^2 Cost = 1000 x £2.25 Cost = £2250	2 x 400W fabric cutters for 200 min Cost of electricity = (200÷60) x 0.4kW x 7.17p x 2 Cost = 19p (2 cutters)	£25.33 £2250.00 .19 £2275.52
MATERIAL PROCESSING	Cutting out skirt pattern for each of 4 sizes	1 operative @ £11.50 per hour for 110 mins Cost = (110÷60) x £11.50 Cost = £21.08	—	1 x 2kW pattern cutting machines for 110 min Cost of electricity = (110÷60) x 2kW x 7.17p Cost = 26p	£21.08 .26 £21.34
ASSEMBLY	Machine-stitching skirt pieces together	4 operatives @ £4.10 per hour for 1005 min Cost = (1005÷60) x £4.10 x 4 Cost = £274.70 (4 ops)	4 cotton reels @ 30p = £1.20 800 buttons @ 1p = £8.00 800 zips @ 60p = £480 Cost = £489.20	4 x 1.5kW sewing machines for 1005 min Cost of electricity = (1005÷60) x 1.5kW x 7.17p x 4 Cost = £7.21 (4 machines)	£274.70 £489.20 £7.21 £771.11
FINISHING	Labelling skirt	4 operatives @ £4.10 per hour for 105 mins Cost = (105÷60) x £4.10 x 4 Cost = £28.70 (4 ops)	800 labels @ 1.5p = £12	4 x 1.5kW sewing machines for 105 min Cost of electricity = (105÷60) x 1.5kW x 7.17p x 4 Cost = 75p (4 machines)	£28.70 £12.00 .75 £41.45
FINISHING	Ironing skirt	2 operatives @ £3.45 per hour for 405 mins Cost = (405÷60) x £3.45 x 2 Cost = £46.58 (2 ops)	—	2 x 1.kW irons for 405 min Cost of electricity = (405÷60) x 1kW x 7.17p x 2 Cost = 97p (2 irons)	£46.58 .97 £47.55
PACKAGING	Packaging skirt	2 operatives @ £3.45 per hour for 80 mins Cost = (80÷60) x £3.45 x 2 Cost = £9.20 (2 ops)	20 boxes @ £1.50 = £30	—	£9.20 £30.00 £39.20
DIRECT COSTS FOR 800 SKIRTS =					£3196.17

Calculation of direct costs for the 800 skirts

Now that we have calculated the direct costs of manufacturing 800 skirts, we can determine the direct costs of manufacturing 1 skirt.

Direct costs of manufacturing 800 skirts = £3196.17
Therefore direct costs of manufacturing 1 skirt
= £3196.17 ÷ 800 = £4.00.

This task is carried out in conjunction with tasks 40, 41, and 42

DIRECT COSTS

- human resources
- material resources
- services resources

TASK 39

Calculation of the direct costs for each key stage of production in the manufacture of a product

Use all the resources available to you, particularly task 38.
Calculate the direct costs for each key stage of production needed to manufacture both your selected products to their agreed outputs using task 38.

For each of your products, carry out the following:

1 Total the direct costs for each key stage of production to find the total direct costs to manufacture your product to its agreed product output.
2 Divide the total direct costs by the agreed product output to determine the total direct costs of one product.

Keep both sets of calculations in your portfolio.

These will form part of the total costs of manufacturing each of your products, which will be put together in task 41.

3.9 INDIRECT COSTS

INDIRECT COSTS

- management
- administration
- marketing
- labour
- materials
- expenses

The indirect costs, sometimes known as overheads or capital costs, are those costs which cannot be directly attributed to a particular stage of production.

Indirect costs can be either fixed or variable costs.

Fixed costs are independent of the work being carried out in the factory, and include rent and rates.

Variable costs fluctuate with the amount of work being carried out and include telephone and administrative costs.

Let's begin by looking at management costs.

➤ Management costs

These costs include the salaries of senior managers, the personnel department, and secretaries.

➤ Administration costs

These costs include the salaries of clerks, typists, and general office staff, and the cost of using telephones, facsimile, photocopiers, and the postal service.

➤ Marketing

These costs include the salaries of the sales and marketing personnel, and the promotion of the product.

➤ Labour

These costs include the salaries of maintenance personnel, caretakers, security, and cleaners.

➤ Materials

These costs include stationery and cleaning materials.

➤ Expenses

These costs include rent, rates, heating, lighting, depreciation, insurance, bank loans, training, and carriers.

Companies differ on how they determine the indirect costs involved in manufacturing a product. One popular method is to multiply the human resource direct costs by a scaling factor, normally:

$$0.8 \times \text{human resource costs}$$

Let's calculate the indirect costs for each production stage in the manufacture of the 800 skirts.

TYPE OF PRODUCTION STAGE	PRODUCTION STAGE	HUMAN RESOURCE DIRECT COSTS	INDIRECT COSTS (0.8 x DIRECT COSTS)
MATERIAL PREPARATION	Cutting cotton fabric to pre-determined lengths	£25.33	£20.26
MATERIAL PROCESSING	Cutting out skirt pattern for each of 4 sizes	£21.08	£16.86
ASSEMBLY	Machine-stitching skirt pieces together	£274.70	£219.76
FINISHING	Labelling skirt	£28.70	£22.96
FINISHING	Ironing skirt	£46.58	£37.26
PACKAGING	Packaging skirt	£9.20	£7.36
INDIRECT COSTS FOR **800** SKIRTS			£324.46

Calculation of indirect costs for the 800 skirts

Now that we have calculated the indirect costs of manufacturing 800 skirts, we can determine the indirect costs of manufacturing one skirt.

Indirect costs of manufacturing 800 skirts = £324.46

Therefore indirect costs of manufacturing 1 skirt = £324.46 ÷ 800 = 41p.

This task is carried out in conjunction with tasks 39, 41, and 42

TASK 40

Calculation of the indirect costs for each key stage of production in the manufacture of a product

Use all the resources available to you, particularly task 39.

Calculate the indirect costs for each key stage of production needed to manufacture both your selected products to their agreed outputs using task 39.

Assume the following at each key stage of production:

> Indirect costs = 0.8 x human resource costs

For each of your products, carry out the following:

1 Total the indirect costs for each key stage of production to find the total indirect costs to manufacture your product to its agreed product output.
2 Divide the total indirect costs by the agreed product output to determine the total indirect costs of one product.

Keep both sets of calculations in your portfolio.

These will form part of the total costs of manufacturing each of your products, which will be put together in task 41.

3.10 TOTAL COST OF MANUFACTURING A PRODUCT

The total cost of manufacturing a product is the sum of the direct costs and indirect costs, ie:

> Total cost of   Direct costs of   Indirect costs of
> manufacturing = manufacturing + manufacturing
> a product   a product   a product

Using this formula, let's calculate the total cost of manufacturing the 800 skirts and hence one skirt.

Direct costs of manufacturing 800 skirts =   £3 196.17
Indirect costs of manufacturing 800 skirts =   £324.46

Total costs of manufacturing 800 skirts =   £3 520.63

Direct costs of manufacturing 1 skirt =   £4.00
Indirect costs of manufacturing 1 skirt =   £0.41

Total costs of manufacturing 1 skirt =   £4.41

The skirt has cost £4.41 to manufacture but at each stage of the production value has been added to the product. This increase in value allows the manufacturer to make a profit and the skirt may be sold to retailers at, say, £7.

This task is carried out in conjunction with tasks 39, 40, and 42

TASK 41

Calculation of the total costs of manufacturing a product

Use all the resources available to you, particularly tasks 39 and 40.

Calculate the total costs of manufacturing both your selected products to their agreed outputs and as individual items.

For each of your products carry out the following calculations:

Total cost of manufacturing agreed product output	=	Total direct costs to manufacture agreed product output	+	Total indirect costs to manufacture agreed product output

Total cost of manufacturing one product	=	Total direct costs of manufacturing one product	+	Total indirect costs of manufacturing one product

Keep the calculations for both products in your portfolio.

3.11 EFFECTS OF CHANGING THE SCALE OF PRODUCTION ON THE COSTS OF A PRODUCT

EFFECTS
• bulk purchasing
• production down-time
• labour
• additional production

SCALE OF PRODUCTION
• continuous flow or line
• repetitive batch
• small batch or job

The following table shows the general effects of changing the scale of production on the costs of a product.

SCALE OF PRO-DUCTION \ EFFECTS ON	BULK PURCHASING	PRODUCTION DOWN-TIME	LABOUR	ADDITIONAL PRODUCTION
CONTINUOUS FLOW OR LINE	CHANGING TO REPETITIVE BATCH — Reduction in bulk purchasing will increase production costs	CHANGING TO REPETITIVE BATCH — Increase in production down-time will increase production costs	CHANGING TO REPETITIVE BATCH — Decrease in labour may reduce production costs	CHANGING TO REPETITIVE BATCH — Will not be able to take on additional production
	CHANGING TO SMALL BATCH OR JOB — Major reduction in bulk purchasing will increase production costs	CHANGING TO SMALL BATCH OR JOB — Major increase in production down-time will increase production costs	CHANGING TO SMALL BATCH OR JOB — Major decrease in labour may reduce production costs	CHANGING TO SMALL BATCH OR JOB — Will not be able to take on additional production
REPETITIVE BATCH	CHANGING TO CONTINUOUS FLOW OR LINE — Increase in bulk purchasing will reduce production costs	CHANGING TO CONTINUOUS FLOW OR LINE — Decrease in production down-time will reduce production costs	CHANGING TO CONTINUOUS FLOW OR LINE — Increase in labour may increase production costs	CHANGING TO CONTINUOUS FLOW OR LINE — Will be able to take on additional production and may reduce production costs
	CHANGING TO SMALL BATCH OR JOB — Decrease in bulk purchasing may increase production costs	CHANGING TO SMALL BATCH OR JOB — Increase in production down-time may increase production costs	CHANGING TO SMALL BATCH OR JOB — Decrease in labour may reduce production costs	CHANGING TO SMALL BATCH OR JOB — Will not be able to take on additional production
SMALL BATCH OR JOB	CHANGING TO CONTINUOUS FLOW OR LINE — Major increase in bulk purchasing will reduce production costs	CHANGING TO CONTINUOUS FLOW OR LINE — Major reduction in production down-time will reduce production costs	CHANGING TO CONTINUOUS FLOW OR LINE — Increase in labour may increase production costs	CHANGING TO CONTINUOUS FLOW OR LINE — Will be able to take on additional production and may reduce production costs
	CHANGING TO REPETITIVE BATCH — Increase in bulk purchasing may reduce production costs	CHANGING TO REPETITIVE BATCH — Reduction in production down-time may decrease production costs	CHANGING TO REPETITIVE BATCH — Increase in labour may increase production costs	CHANGING TO REPETITIVE BATCH — Will be able to take on additional production and may reduce production costs

This task is carried out in conjunction with tasks 39, 40, and 41

EFFECTS

- bulk purchasing
- production down-time
- labour
- additional production

SCALE OF PRODUCTION

- continuous flow or line
- repetitive batch
- small batch or job

TASK 42

Explaining the effects of changing the scale of production on the costs of a product

To carry out this task, use all the resources available to you – such as the library, magazines, journals, teachers, tutors, local companies, and 'yellow pages'.

Select one of your two products.

In broad terms explain the effects of changing the scale of production on the costs of your product.

To carry out this task, follow these steps:

1 Initially consider your product to be manufactured by a continuous flow or line production system.

2 Suppose you have to change to a repetitive batch production system. Explain what the listed effects would have on your production costs.

3 Now suppose you have to change from a continuous flow or line production system to a small batch or job production system. Explain what the listed effects would have on your production costs.

4 Now consider your product to be manufactured by a repetitive batch production system.

In turn, change to the other two types of production system and explain what the listed effects would have on the production costs.

5 Finally, consider your product to be manufactured by a small batch or job production system.

In turn, change to the other two types of production system and explain what the listed effects would have on the production costs.

Bring together tasks 39–42 inclusive to produce a report titled: 'Calculate the cost of a product'.

Keep the report in your portfolio.

3.12 QUALITY ASSURANCE SYSTEMS

A quality assurance system is one whose activities and functions are mainly concerned with the attainment of quality. To achieve this, the quality assurance system needs to be an integral part of the way the organization functions, involved from the design through the manufacture of the product to its after-sales service.

Quality assurance builds in quality at all stages of the process and reduces the need for quality control on the finished product. British Standards (BS) 5750 and International Standards Organization (ISO) 9000 provide the quality standards which should be applied by everyone in the organization.

Quality control plays an important role within quality assurance and refers to the operational techniques and activities used to measure, record and maintain the standards of production. Quality control is achieved by establishing the quality standards required by the customer and then planning to achieve these standards.

As the product is manufactured it is inspected and, if necessary, corrective action is taken to ensure that the expected quality standards are met. Many manufacturing organizations are fully committed to manufacturing products 'right first time' since defective products cost them money.

RIGHT FIRST TIME is directly concerned with getting the quality of a product right first time. Manufacturing a quality product is the result of quality manufacturing and quality control. Quality cannot be put into the product by inspecting it: the product has to be made correctly from the start, ie 'right first time'.

Let's now turn our attention to the key factors of quality assurance system.

Let's begin by looking at the organization of the workforce in a quality assurance system.

> **DEFINITION**
> Defective product: a product which does not meet the required quality standards.

> KEY FACTORS OF A QUALITY ASSURANCE SYSTEM
>
> - organization of workforce
> - control of design
> - control of production systems
> - manufacture to specification
> - standards

➤ Organization of workforce

The organization of the workforce is a very important factor in the success of a quality assurance system. The way the workforce is organized has a considerable effect on how they perform, and of course this has a direct bearing on the quality of the product being manufactured. The workforce must be given roles that are displayed on an organization chart for everyone to see. You first met organization charts in section 1.11.

The managers in an organization must be given clearly allocated responsibilities towards quality, and the Chief Executive must take overall responsibility for quality.

In many organizations, teams are used as a method of actively involving and motivating the workforce in working towards a continuous and company-wide quality improvement programme. The workforce need to feel they are a part of the whole organization and understand why they are doing things in a particular way. Amongst the most popular types of team are project teams, quality circles and quality improvement teams.

Quality Control

(a) In the brewing industry

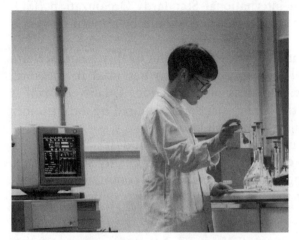

(b) In the chemical industry

(c) In the food industry

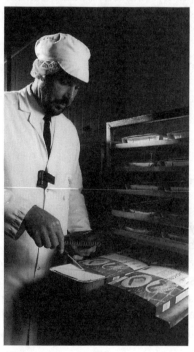

(d) In the food industry

(e) In the paper industry

➤ Control of design

The control of design in an organization is a very important step towards making sure that the customer requirements are met.
Control of design includes the following activities:

- planning the design and development programme;
- tasking the appropriate design teams;
- working to laid-down codes and procedures;
- review and evaluation of the design.

➤ Control of production systems

The control of production involves a great deal of planning, together with its associated documentation such as:

- production layout;
- machinery and equipment requirements;
- movement of materials and their availability;
- movement of the product;
- procedures for handling substandard products.

➤ Manufacture to specification

The specification is defined as the document that prescribes the requirements with which the product has to conform.

To manufacture to specification, a product must be inspected at various stages of its manufacture such as incoming materials, in-process and finished product.

➤ Standards

Two very important quality management systems are British Standards BS 5750 and International Standards Organization ISO 9000.

◆ BS 5750

BS 5750 was originally developed by the engineering sector and contains a series of standards aimed at specifying quality system requirements to manufacturers.

BS 5750 – Part 1 – 1979
Specification for design, manufacture and installation
This standard states the system to be applied when the technical material requirements are specified in terms of performance or where design has not yet been established.

BS 5750 – Part 2 – 1979
Specification for manufacture and installation
This standard states the system to be applied when the technical material requirements are specified in terms of established design, and where the specified product requirements are inspected and tested during manufacture.

BS 5750 – Part 3 – 1979
Specification for final inspection and test
This standard states the system to be applied when the specified product requirements can be inspected and tested at the final stage of manufacture.

BS 5750 – Parts 4, 5, and 6 are guidelines for parts 1, 2, and 3 respectively.

◆ ISO 9000

ISO 9000 is a set of quality system standards published in 1987 that were based on BS 5750 and have now been adopted by many countries in the European Community and throughout the rest of the world. It consists of five individual standards that are divided into four parts.

The standards cover requirements of a quality system such as:

- establishing and maintaining the system;
- management responsibility of the system;
- design and document control;
- contract review;
- purchasing;
- process control;
- control of purchased materials;
- control of measuring and test equipment;
- control of non-conforming products;
- corrective action;
- quality records;
- audits and reviews;
- training.

This task is carried out in conjunction with tasks 44, 45, 46, 47, and 51

KEY FACTORS

- organization of work-force
- control of design
- control of production systems
- manufacture to specification
- standards

TASK 43

Identify the key factors of a quality assurance system

To carry out this task, use all the resources available to you – such as the library, magazines, journals, teachers, tutors, local companies, and 'yellow pages'.

Identify, in general terms, the key factors of a quality assurance system within a manufacturing organization.

Produce a brief report on the key factors of a quality assurance system. This should include a description of the difference between quality control and quality assurance.

Keep this brief report in your portfolio.

EXAMPLES OF QUALITY
INDICATORS

- weight
- volume
- size
- functionality
- appearance
- taste
- sound
- smell
- touch

DEFINITION
Variable: a product has some property or properties that vary, such as length or weight.

DEFINITION
Attribute: a product has some attribute or attributes that are either 'good' or 'bad', 'acceptable' or 'not acceptable', or 'pass' or 'fail'.

This task is carried out in conjunction with tasks 43, 45, 46, 47, and 51

3.13 FUNCTION OF QUALITY INDICATORS AND CRITICAL CONTROL POINTS

We met quality indicators and critical control points briefly at the start of this chapter and also in Chapter 1.

Let's begin by looking at what quality indicators are.

➤ Quality indicators

A quality indicator is a variable or an attribute of a product that can be measured or assessed respectively. The data obtained from measurement or assessment can be compared with the product's specification to give an indication of its quality.

And now let's look at the function of critical control points.

➤ Critical control points

A critical control point is a point in the manufacture of a product where the product is inspected/monitored to ensure that it is being successfully/correctly manufactured to specification.

There can be several critical control points identified within the manufacture of a product starting with the inspection of the raw materials, through the various production processes to the finished product.

Quality indicators are applied at each critical control point.

Task 44
Describe the function of quality indicators and critical control points

To carry out this task, use all the resources available to you.

Describe, in general terms, the function of quality indicators and control points.

- What is a quality indicator? Give examples of quality indicators.
- What is a critical control point? Give examples of where critical control points would be set up.
- What is the relationship between quality indicators and critical control points?

Produce a brief report on the function of quality indicators and critical control points.

Keep this brief report in your portfolio.

3.14 ROLE OF TESTING AND COMPARISON IN QUALITY CONTROL

Testing and comparison of a product are carried out within quality control in order to assess whether a product conforms to its specification.

The product is tested, very often electronically, for some variable such as volume, length, or weight, and the collected data is then compared with the specified data for that product. Provided the variable is within certain limits, then it is accepted and the product is said to conform to its specification.

A product can also be tested visually for some attribute such as colour, finish, or texture, and then compared with a standard, perhaps using a colour chart or more often one's past experience. Provided the attribute is within certain limits, then it is accepted and the product is said to conform to its specification.

This task is carried out in conjunction with tasks 43, 44, 46, 47, and 51

TASK 45

Explaining the role of testing and comparison in quality control

To carry out this task, use all the resources available to you.

Explain, in general terms, the role of testing and comparison in quality control.

- Why are testing and comparison carried out in quality control?
- What methods of testing are used?
- What methods of comparison are used?

Produce a brief report on the role of testing and comparison in quality control.

Keep this brief report in your portfolio.

3.15 QUALITY CONTROL TECHNIQUES

A product can be inspected and tested at any stage of its manufacture to assess whether it conforms to its specification.

If the batch of products being manufactured is small, then all the products may be inspected and tested, ie 100% inspection. However, if the batch of products is large, then because of the costs and time involved, they may be sampled using a random sampling method. This means that a sample of the whole batch is removed, inspected and tested, and on the results of this inspection the whole batch is accepted or rejected.

Let's begin by looking at the various inspection and testing methods used within quality control.

➤ Inspection and testing methods

INSPECTION AND TESTING
METHODS

- visual
- mechanical
- electrical and electronic
- expert scrutiny
- chemical analysis

Let's start by looking at visual methods.

◆ Visual

The colour, finish, and texture of the manufactured product are inspected visually, sometimes using a magnifying glass or microscope, or using a colour chart to compare with a standard.

◆ Mechanical

Mechanical inspection and testing methods are generally classified into two main types: destructive and non-destructive testing.
First let's look at destructive testing.

◆ Destructive testing

This is carried out in such a way that the product or test piece is destroyed during the test.

The tensile strength of most types of material can be determined. A manufactured part or product, such as a bolt, can be mounted in a tensile test machine and stressed to breaking point to determine its tensile strength.

A sewing thread, such as cotton, or a fibre, such as rayon used in the tyre-cord of a tyre, can be tested for tensile strength and extendibility, ie the amount of stretch before breaking.

Impact testing of many different types of material is used to determine the ability of a material to withstand shock loading. Impact strength is very important for manufactured products such as automobiles, luggage, and safety helmets.

Hardness testing of many types of material is used to determine the ability of a product to withstand indentations and in some cases scratches. Hardness is important for products such as ball and roller bearings. The abrasion resistance of products such as windscreens, floor tiles, and sinks is also very important.

Fatigue testing determines the effects of a force applied to the product. All types of material can fail when repeatedly subjected to stresses. Fatigue testing is important for products such as fishing rods, foam cushions, and turbine blades.

Compressive strength is the ability of a material to resist forces that tend to crush it. It is important for products such as bricks, toys, and furniture.

Another form of destructive testing is carried out when food products are cut open and inspected, tasted and, in some cases, tested for bacteria.

◆ Non-destructive testing

Non-destructive testing is carried out in such a way that the product or test piece is not destroyed during the test. Tests are carried out to ensure that the product is sound and free from defects which could cause failure.

Ultrasonic inspection and dye which penetrates a product are used to inspect for cracks and flaws in large volumes of material, such as railroad wheels.

X-ray inspection is used to detect internal cracks and flaws in many different types of engineering product, for example pressure vessels.

Eddy current inspection is used to detect defects in a product. Eddy currents are caused to flow in the product and any defect will hold up the eddy currents. This can then be detected using a voltmeter.

Dimensional testing is carried out using micrometers to check the accuracy of a manufactured product.

Now let's look at electrical and electronic inspection and testing methods.

◆ Electrical and electronic

Electrical and electronic inspection and testing methods include weighing products using electronic weighing scales which are normally more accurate than traditional scales. These are used on food products which are sold by weight.

The size of a product can be determined by electronic gauging, using a device designed to reject any product outside the specified limits.

The functionality of a product can be determined using measuring instruments such as decibel meters, and test instruments such as automatic test equipment (ATE).

◆ Expert scrutiny

An expert scrutineer is a person who has expert knowledge of the product and inspects and tests products such as musical instruments, stamps, bank notes and coins.

◆ Chemical analysis

Many different types of product have to be inspected and tested by taking samples of the product to a laboratory for chemical analysis. Products such as perfumes and cosmetics, fibreglass, fertilizers, and medicines are chemically analysed. Many products are tested for pH which has a scale from 0 to 14 for measuring acidity or alkalinity, with below 7 for acids, above 7 for alkalis and 7 for neutrals. The pH can be measured using a solution and a paper strip or a pH meter.

Battery acid, soap, and caustic soda have pH readings of 2, 9, and 13 respectively.

And now let's investigate sampling techniques.

➤ Sampling techniques

As we saw earlier in this section inspecting every product is costly and time consuming and so some form of sampling inspection is used whenever possible. This is known as acceptance sampling, because a decision about the quality of a batch of products is made after inspecting and testing a portion of the batch. If the sample of products conforms to the specified quality levels, then the whole batch is accepted; if the sample does not conform to the specified quality levels, then the whole batch is rejected or, in some cases, subjected to further inspection.

Various types of sampling technique are used in the quality control department. Let's start by investigating single sampling.

◆ Single sampling

Only one sample of a batch of products is taken and inspected. If the number of defective products is equal to or less than an acceptance figure, then the whole batch of products is accepted.

For example, we may produce a batch of 1 000 products and take a sample of 40. If two or less of these 40 products are defective, then we accept the whole batch; conversely, if there are more than two defects, then we reject the whole batch.

Now $\frac{2}{40}$ as a percentage is 5% and so we are accepting up to and including 5% defective products. A manufacturer knows what a customer will accept and this is known as the Acceptable Quality Level (AQL).

If an Acceptable Quality Level is 5%, then the customer will not accept 6% defective products; this is known as Lot Tolerance Percentage Defective (LTPD).

If we manufacture a batch with an unacceptable level of defective products, then we may have to carry out one of the following actions:

- carry out a 100% screening and rectify the defects;
- sell them to a customer who will accept them;
- offer them to a customer at a discounted price;
- carry out another single sample inspection and confirm or otherwise whether there was an unacceptable level of defective products.

It is extremely important to ensure that the size of the sample is right. Small samples can give misleading results and large samples are expensive and time consuming to check.

Another commonly used sampling technique is known as random sampling.

◆ Random sampling

The first product manufactured in a batch is inspected to make sure it conforms to the specified quality standards; then other products are selected randomly for inspection. Random sampling can suffer from

DEFINITION
Acceptable Quality Level (AQL): the highest percentage level of defects in a batch which a customer will regularly accept.

DEFINITION
Lot Tolerance Percentage Defective (LTPD): the lowest percentage of defective products which a customer will not accept.

sampling error which occurs when the selected sample fails to give a true indication of the condition of the products. The larger the sample, the less likely that sampling error will occur.

Let's now look at 100% or continuous sampling.

◆ Continuous sampling

In this form of sampling, every product is inspected as it is produced. As previously mentioned, this is expensive and time consuming, but it does ensure that a quality product is manufactured.

And finally let's investigate multiple sampling.

◆ Multiple sampling

Multiple sampling is a more sophisticated form of single sampling. There are three possible decisions that can be taken after the inspection of a sample of the products from a batch, ie accept the whole batch, reject it, or take a further sample from the batch. If a further sample is taken, then the results are added to those of the first sample and an accept or reject decision about the whole batch is made. This has the advantage that inspectors are reluctant to reject entire batches that only just fail.

This task is carried out in conjunction with tasks 43, 44, 45, 47, and 51

INSPECTION AND TESTING

- visual
- mechanical
- electronic and electrical
- expert scrutiny
- chemical analysis

SAMPLING

- single
- random
- continuous
- multiple

TASK 46

Identify and describe quality control techniques

To carry out this task, use all the resources available to you.

Identify and describe, in general terms, the quality control techniques of inspection and testing, and sampling.

Produce a brief report on quality control techniques.

Keep this report in your portfolio.

3.16 USE OF TEST AND COMPARISON DATA

We met the role of testing and comparison in section 3.14, where we found that it is carried out within quality control in order to assess whether a product conforms to its specification. In this section we are going to describe the use of test and comparison data.

Let's begin by investigating batch to batch conformity to specification.

▶ Batch to batch conformity to specification

Batch to batch conformity to specification is carried out using statistical process control, which is basically a measurement tool.

Let's begin by looking at an example.

Suppose a machine is required to produce a batch of 800 metal discs of radius 8 ± 0.03mm. (±0.03mm is known as the tolerance.) This means that the metal disc is acceptable if its radius lies between 7.97mm and 8.03mm, and unacceptable if it lies outside those limits.

A sample of 45 metal discs is inspected and the results are as follows:

EXAMPLE

DEFINITION
Tolerance: permissible deviation from the nominal value.

Radius (mm)	Number of metal discs
7.98	6
7.99	10
8.00	15
8.01	9
8.02	5
Total	45

Inspection is carried out using control of variables, ie variation in the radius measurement. From these results we can produce a histogram.

DEFINITION
Action line: limits specified by the customer.

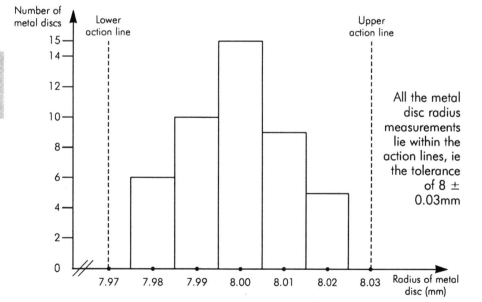

Histogram of the sample of 45 discs

Suppose the results of the next sample inspection are as follows:

Radius (mm)	Number of metal discs
7.99	6
8.00	9
8.01	15
8.02	10
8.03	—
8.04	5
Total	45

From these results we can produce the following histogram.

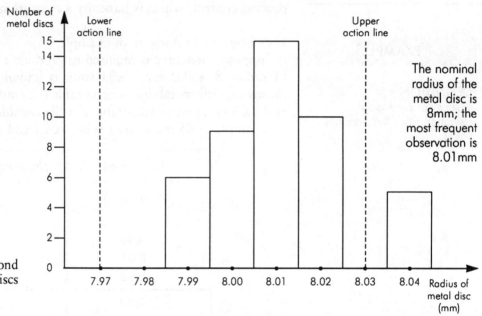

The nominal radius of the metal disc is 8mm; the most frequent observation is 8.01mm

Histogram of the second sample of 45 discs

To overcome this problem, you could ask the machine operator to centre the work at just below 8mm.

Five metal discs are over the limit, ie $\frac{5}{45}$ which is 11.1%.

In the sample of 45 metal discs out of a batch of 800, five of the discs are defective. We cannot assume the remaining 755 will have 11.1% defective discs, and thus we need to take more samples to get a true indication.

With process operations that produce variable outputs, under normal working conditions, we need a technique to be able to interpret whether the results of a sample inspection of a batch of products are as expected or not. The technique uses **control charts**.

There are two main types of control chart: average and range.
The average value of the variable measured in each sample is plotted on the average control chart.
The range value (difference between the greatest and least values) of

DEFINITION
Control chart: a chart designed to inform the operator whether a process operation is continuing to perform as it did when the control limits were set initially.

the variable measured in each sample is plotted on the range control chart.

Let's look at an average control chart which could be used to monitor the batch to batch conformity of the metal disc specification.

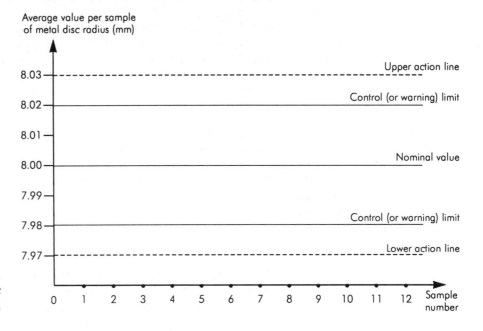

Average control chart for sampling 45 items out of batch sizes of 800

The upper and lower action lines are the limits set by the specification of the product, ie the tolerance.

The nominal value is the customer's preferred value of the variable being measured.

The control (or warning) limits are calculated so that they are only to be exceeded once in every 40 times under random conditions.

Let's work out the average values of the first and second sample inspections of the metal discs that we introduced earlier.

First sample inspection
Average value =
$$\frac{(6 \times 7.98) + (10 \times 7.99) + (15 \times 8.00) + (9 \times 8.01) + (5 \times 8.02)}{40}$$
Average value = 8.00 mm (to 2 decimal places)

Second sample inspection
Average value =
$$\frac{(6 \times 7.99) + (9 \times 8.00) + (15 \times 8.01) + (10 \times 8.02) + (5 \times 8.04)}{40}$$
Average value = 8.01 mm (to 2 decimal places)

Suppose that the next 10 sample inspections produce the following average values:
8.00mm, 8.01mm, 7.99mm, 8.00mm, 8.00mm, 8.01mm, 8.01mm, 8.01mm, 7.99mm, 7.99mm.

Let's plot these 12 sample average values on an average control chart:

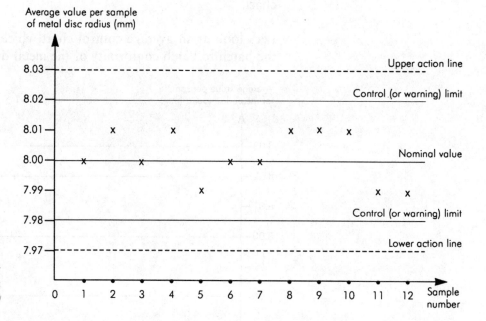

Average control chart for 12 samples of 45 items from batch sizes of 800

The control limit is calculated so that it is only expected to be exceeded once in every 40 times. This means that if we take 40 samples, then we expect the control limit to be exceeded only once if variations occurring are random. If there are more than one exceeding the control limits, then there is likely to be a fault in the system and it needs to be rectified.
In the above example, we can see that the average value of the samples from each batch conforms to the specification.

Let's now investigate the range control chart.
Here is a range control chart which could be used to monitor the batch to batch conformity of the metal disc specification:

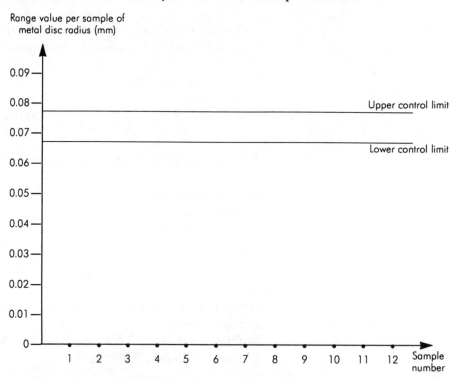

Range control chart for sampling 45 items out of batch sizes of 800

The upper control limit is calculated such that it is only expected to be exceeded once in every thousand times.

The lower control limit is calculated such that it is only expected to be exceeded once in every 40 times. This means that if we take 40 samples, then we expect the lower control limit to be exceeded only once if variations occurring are random. If there are more than one exceeding the lower control limit, then there is likely to be a fault in the system, and it needs to be rectified.

Let's work out the range values of the first and second sample inspections of the metal discs that we introduced earlier.

First sample inspection
Range value = largest value–smallest value
Range value = 8.02mm–7.98mm
Range value = 0.04mm

Second sample inspection
Range value = 8.03mm–7.99mm
Range value = 0.04mm

Suppose that the next 10 sample inspections produce the following range values:
0.03mm, 0.04mm, 0.03mm, 0.03mm, 0.04mm, 0.02mm, 0.03mm, 0.02mm, 0.04mm, 0.03mm

Let's plot these 12 sample range values on a range control chart.

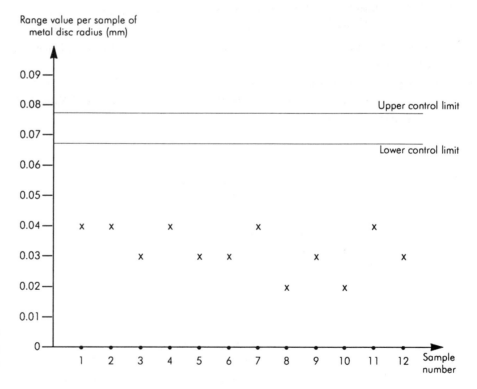

Range control chart for 12 samples of 45 items from batch sizes of 800

In the above example, we can see that the range value of the samples from each batch conforms to the specification.

Let's look at another use of the test and comparison data.

➤ Monitoring for improvement

Test and comparison data can be used to monitor improvements in the product, for example when more efficient use of materials and different work practices are introduced.

➤ Calibration and checking of control and monitoring equipment

Control and monitoring equipment that is used to inspect, measure, and test must be accurate consistently in order to obtain reliable test and comparison data. The equipment should be calibrated and checked by comparing with reference standards in-house, or at a specialist laboratory.

This task is carried out in conjunction with tasks 43, 44, 45, 46, and 51

USE
- batch to batch conformity to specification
- monitoring for improvement
- calibration and checking of control and monitoring equipment

TASK 47

Describe the use of test and comparison data

To carry out this task, use all the resources available to you.

Describe, in general terms, the use of test and comparison data.

Monitoring for improvement could include reducing costs through more efficient use of materials and different working practices as well as improving environmental conditions.

Calibration and checking of control and monitoring equipment is essential to ensure that data is accurate consistently.

Produce a brief report on the use of test and comparison data.

Keep this report in your portfolio.

This is a stand-alone task and will not form part of any other report

TASK 48

Sampling

Complete the following task.

Five batches of plastic discs are produced by a plastics manufacturer. Each batch consists of 600 discs and a sample of 30 discs is taken from each batch for inspection. Each plastic disc is specified to have a radius of 5 ± 0.1cm and a finite thickness.

The results of the five sample inspections are as shown opposite.

SAMPLE INSPECTION 1		SAMPLE INSPECTION 2		SAMPLE INSPECTION 3		SAMPLE INSPECTION 4		SAMPLE INSPECTION 5	
RADIUS (cm)	NO OF DISCS	RADIUS (cm)	NO OF DISCS	RADIUS (cm)	NO OF DISCS	RADIUS (cm)	NO OF DISCS	RADIUS (cm)	NO OF DISCS
4.88	–	4.88	2	4.88	–	4.88	–	4.88	–
4.90	–	4.90	2	4.90	–	4.90	–	4.90	–
4.92	–	4.92	4	4.92	1	4.92	–	4.92	–
4.94	–	4.94	6	4.94	2	4.94	–	4.94	–
4.96	2	4.96	12	4.96	2	4.96	15	4.96	–
4.98	3	4.98	2	4.98	3	4.98	8	4.98	10
5.00	10	5.00	2	5.00	18	5.00	4	5.00	10
5.02	6	5.02	–	5.02	2	5.02	2	5.02	10
5.04	3	5.04	–	5.04	2	5.04	1	5.04	–
5.06	2	5.06	–	5.06	–	5.06	–	5.06	–
5.08	2	5.08	–	5.08	–	5.08	–	5.08	–
5.10	1	5.10	–	5.10	–	5.10	–	5.10	–
5.12	1	5.12	–	5.12	–	5.12	–	5.12	–

(a) Calculate and plot the average values of each sample on the following average control chart.

Average value of sample 1 =   Average value of sample 2 =

Average value of sample 3 =   Average value of sample 4 =

Average value of sample 5 =

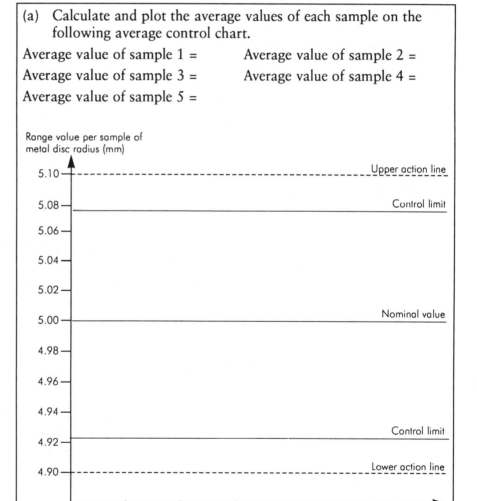

Range value per sample of metal disc radius (mm)

Average control chart for sampling 30 items out of batch sizes of 600

Are all the average values of the samples within the control limits? Comment on the results.

(b) Calculate and plot the range values of each sample on the following range control chart.

Range value of sample 1 =   Range value of sample 2 =

Range value of sample 3 =   Range value of sample 4 =

Range value of sample 5 =

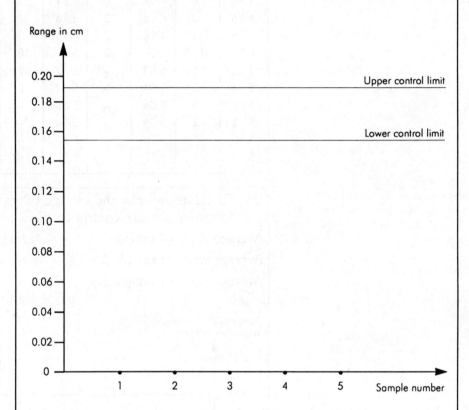

Range control chart for sampling 30 items out of batch sizes of 600

Are all the range values of the samples within the control limits?
Comment on the results.
What improvements would you make to the process?

(c) Produce a histogram of each sample inspection (using a computer statistics package).

Make suitable comments about each histogram and, if necessary, make realistic improvements to the process producing the plastic discs.

TASK 49

Inspecting a sample of items

Complete the following task.

Obtain at least four packs of eggs of the same size number, ie size 0 to 7, with the same packing station identification number and week number.

Egg sizes				
		Size 0	≤	75g
70g	≤	Size 1	<	75g
65g	≤	Size 2	<	70g
60g	≤	Size 3	<	65g
55g	≤	Size 4	<	60g
50g	≤	Size 5	<	55g
45g	≤	Size 6	<	50g
		Size 7	<	45g

(a) Accurately weigh each egg to the nearest gram.

(b) Produce a table of results.

(c) Draw a histogram of the results using 1 gram intervals.

(d) Do all the eggs weigh within the specified tolerances?

(e) If not, then suggest reasons why this is so.

(f) Investigate how eggs are packed.

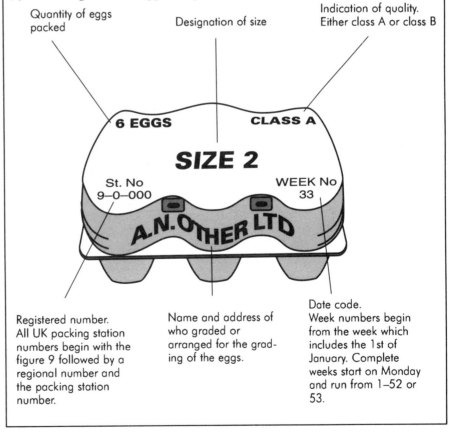

Quantity of eggs packed

Designation of size

Indication of quality. Either class A or class B

6 EGGS   CLASS A

SIZE 2

St. No 9–0–000   WEEK No 33

A.N.OTHER LTD

Registered number. All UK packing station numbers begin with the figure 9 followed by a regional number and the packing station number.

Name and address of who graded or arranged for the grading of the eggs.

Date code. Week numbers begin from the week which includes the 1st of January. Complete weeks start on Monday and run from 1–52 or 53.

Pre-pack markings of egg packs

This is a stand-alone task and will not form part of any other report

TASK 50

Investigating a sample of items

Obtain a pack of resistors of 5% or 10% tolerance, of a known nominal value such as 1 000 ohms.

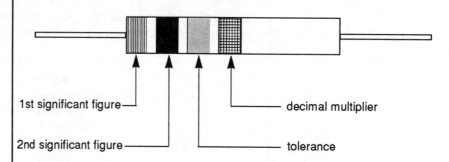

Resistor colour code

Colour	Significant figure	Decimal multiplier	Tolerance (per cent)
silver		0.01	10
gold		0.1	5
black	0	1	
brown	1	10	
red	2	10^2	
orange	3	10^3	
yellow	4	10^4	
green	5	10^5	
blue	6	10^6	
violet	7	10^7	
grey	8	10^8	
white	9	10^9	

For example, a 1 000 ohm resistor with a 10% tolerance would have the following colour code:

$10 \times 10^2 = 1\ 000$ ohms

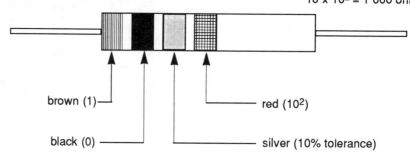

brown (1)

black (0)

red (10^2)

silver (10% tolerance)

(a) Accurately measure each resistor.
(b) Produce a table of results.
(c) Draw a histogram of the results using suitable intervals.
(d) Are all the resistors within the specified tolerance?
(e) Investigate how resistors are produced.

3.17 APPLICATION OF QUALITY CONTROL

APPLICATION OF QUALITY CONTROL

• quality indicators
• critical control points

QUALITY INDICATORS

• weight
• volume
• size
• functionality
• appearance
• taste
• sound
• smell
• touch

EXAMPLES

The application of quality control to a given product is mainly in terms of quality indicators and critical control points.

As we have already seen, quality indicators are applied at critical control points. Before we look at an example of applying quality control to a given product, let's look at some general examples of quality indicators.

➤ Quality indicators

◆ Weight

This is a quality indicator which indicates if the weight of the product is within a stated tolerance.
Example: an egg size No 2 must weigh greater than or equal to 65g and less than 70g, ie $65g \leq$ egg size No 2 $< 70g$.

◆ Volume

This is a quality indicator which indicates if the volume of the product is within a stated tolerance.
Example: a litre of a fluid substance may have a volume greater than 0.98 litres and less than 1.02 litres, ie 1 ± 0.02 litres.

◆ Size

This is a quality indicator which indicates if the size of the product is within a stated tolerance. Size includes length, diameter, thickness, and area.
Example: the nominal length of a manufactured product is 64cm and is specified to be greater than 63.95cm and less than 64.05cm, ie 64 ± 0.05cm.
Example: the nominal thickness of a manufactured part is 50mm and is specified to be greater than 48mm and less than 52mm, ie 50 ± 2mm.

◆ Functionality

This is a quality indicator which indicates if the product is functioning correctly.
Example: the nominal maximum power rating of a hi-fi loudspeaker is 16 watts and is specified to be greater than 15.5 watts and less than 16.5 watts, ie 16 ± 0.5 watts.
Example: whether the on/off switch of a torch functions correctly or not.

◆ Appearance

This is a quality indicator which indicates if the product appears to be acceptable or not. It includes the finish, colour, and texture of the product.
Example: the surface of a piece of furniture is either scratch-free or it is scratched.

◆ Taste

This is a quality indicator which indicates if the taste of the product is acceptable or not.
Example: a wine taster will taste a sample of the wine produced by a vineyard in order to find out whether or not the taste of the product is acceptable.

◆ Sound

This is a quality indicator which indicates if the sound produced by the product is audible enough or within specified tolerances.
Example: a smoke alarm is tested for correct audibility using the test button; a musical instrument is tested for its tone and pitch.

◆ Smell

This is a quality indicator which indicates if the smell of the product is acceptable or not.
Example: the aroma of a perfume produced by a cosmetic company will be tested in order to indicate if the smell of the product is acceptable or not.

◆ Touch

This is a quality indicator which indicates if the touch (feel) of the product is acceptable or not.
Example: the 'feel' of the products manufactured by a leather goods company will be tested in order to indicate if the finish of the product is acceptable or not.

Now let's investigate the application of quality control to the manufacture of the metal tray that we saw earlier in the chapter.

QUALITY INDICATOR

1. The surface of the sheet metal is checked for pitting and discolorations. *(Appearance)*

• Critical control point 1

CUTTING METAL | **Material preparation**

FOLDING METAL | **Processing**

2. The dimensions of each tray are measured. *(Size)*

•Critical control point 2

SPOT-WELDING | **Assembly**

WIRE BRUSHING | **Finishing**

GALVANIZING | **Finishing**

3. The quality of the finished product is checked. *(Appearance and Functionality)*

•Critical control point 3

PACKAGING | **Packaging**

Key production stages in the manufacture of a metal tray

Now let's investigate the application of quality control to the manufacture of a sports skirt.

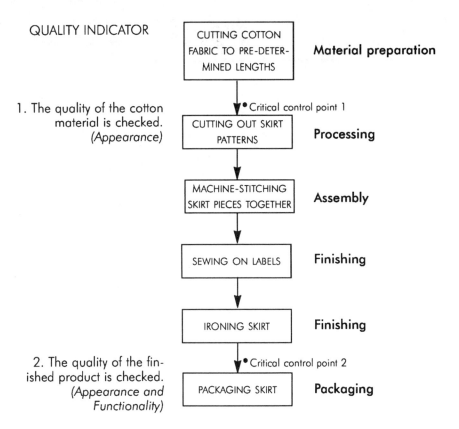

QUALITY INDICATOR

CUTTING COTTON FABRIC TO PRE-DETERMINED LENGTHS — **Material preparation**

1. The quality of the cotton material is checked. (Appearance)

●Critical control point 1

CUTTING OUT SKIRT PATTERNS — **Processing**

MACHINE-STITCHING SKIRT PIECES TOGETHER — **Assembly**

SEWING ON LABELS — **Finishing**

IRONING SKIRT — **Finishing**

2. The quality of the finished product is checked. (Appearance and Functionality)

●Critical control point 2

PACKAGING SKIRT — **Packaging**

Key production stages of the manufacture of a sports skirt

This task is carried out in conjunction with tasks 43, 44, 45, 46, and 47

QUALITY CONTROL
- quality indicators
- critical control points

KEY STAGES OF PRODUCTION
- material preparation
- processing
- assembly
- finishing
- packaging

TASK 51

An investigation into how quality control is applied to a given product at the key stages of production

To carry out this investigation, use all the resources available to you, particularly task 38.

In task 38, you produced production plans for both of your selected products. Select one of the products for this task.

In the production plan for this product, you produced a flow diagram of the key stages of production and you also identified the critical control points and the quality indicators that would be used at these points.

Apply quality control to your product at the key stages of production.

To carry out this task, use the flow diagram of the key stages of production, identify the critical control points on the diagram, and describe the quality indicators to be used at these points.

Bring together tasks 43, 44, 45, 46, 47, and 51 to produce a report titled: 'Investigate quality assurance'.

Keep the report in your portfolio.

4 · MANUFACTURING PRODUCTS

4.1 KEY CHARACTERISTICS OF MATERIALS

Materials are normally selected for the manufacture of a product with the following considerations in mind:

- the conditions to which the product will be subjected;
- the process to which the materials will be subjected;
- the cost of the materials and processing.

Let's investigate the key characteristics of materials that are selected for the manufacturing of products.

We'll begin by looking at the mechanical properties of materials.

KEY CHARACTERISTICS OF MATERIALS

- mechanical properties
- electrical properties
- thermal properties
- physical properties
- composition
- resistance to degradation

➤ Mechanical properties

◆ Stiffness

Stiffness is the ability of a material to resist bending. Fibre composite products such as rackets and skis and many wooden products have this property.

◆ Elasticity

Elasticity is the ability of a material to stretch and return to its original shape. Examples of this are nylon and wool products.

◆ Hardness

Hardness is the ability of a material to resist abrasion or indentation. The hardness of materials ranges from soft talc to diamonds, which are the hardest of all materials.

Property	Ceramics	Metals	Polymers
Hardness	High	Medium	Low

◆ Strength

Strength is the ability of a material to resist applied forces without breaking. Ceramic materials in general have the higher strengths, with diamond having the highest of all. Many composite materials and metals have similar strengths.

◆ Brittleness

A fragile material is said to be brittle, ie hard and easily broken. Products such as glass, pottery, and cast iron are examples of brittle materials.

◆ Ductility

A ductile material is one that suffers permanent deformation before breaking. Metals are very ductile materials. Products such as cars, aluminium cans, and electrical wiring are manufactured from ductile materials.

Now let's look at the electrical properties of materials.

➤ Electrical properties

◆ Electrical conductivity

Electrical conductivity is a measure of the ability of a material to conduct electricity. For example, copper is used as a conductor in toasters, electric fires, and electrical cables.

Property	Ceramics	Metals	Polymers
Electrical conductivity	Low	High	Low

Now let's consider the thermal properties of materials.

➤ Thermal properties

◆ Thermal conductivity

Thermal conductivity is a measure of the ability of a material to conduct heat. For example, aluminium and copper are used in the manufacture of many of the products used in cooking.

Property	Ceramics	Metals	Polymers
Thermal conductivity	Medium	High	Low

➤ Physical properties

◆ Flavour

Taste: flavour is a quality of food or liquid that can be experienced by the mouth. Flavouring is a substance used to give flavour to a food or liquid. Natural and artificial essences are flavourings, which include vanilla essences and almond oil.

◆ Colour

Colour is a sensation produced on the eye by light waves of different lengths. The three primary colours are red, green, and blue.

The colour of the materials used in the manufacture of a product can be very important, particularly in the textiles and clothing industries.

Colouring is a food additive used to alter or improve the colour of processed foods. Colourings include natural colours such as carotene and caramel, and artificial colours such as amaranth and tartrazine, which are made from petrochemicals.

◆ Density

Density is a measure of the compactness of a material.

$$\text{density } \rho = \frac{\text{mass}}{\text{volume}} \text{ kg m}^{-3}$$

Here are a few densities of some common materials.

Material	Density (kg m^{-3})
Lead	11 000
Aluminium	2 600
Oak	650
Balsa wood	200

Many products are made from materials which have a high strength with as low a density as possible. Two examples of materials with these properties are titanium and magnesium alloys.

Now let's look at the composition of materials.

➤ Composition

◆ Composites

Composite materials are composed of at least two materials and they are designed to have properties which are different and sometimes superior to those of the individual materials. It is commonplace nowadays to find lengths of asbestos, glass fibre, carbon fibre, or steel composed in materials such as plastic, concrete, or epoxy resins.

Reinforced concrete is composed of cement and sand reinforced with steel wire.

A composite material of carbon fibre reinforced epoxy resin is used in the manufacture of rackets, and glass fibre reinforced epoxy resin is used in the manufacture of skis.

Wood is regarded as an advanced composite material consisting of layers of cellulose fibres spirally wound in a polymer known as lignin.

◆ Mixture

A mixture is the combining or blending of ingredients into a single mass, such as the mixing of flour, eggs and milk into a batter, or sand and cement into concrete. Blending involves mixing different grades or varieties of tea, tobacco, coffee, and whisky to produce a distinctive taste and aroma.

◆ Alloy

An alloy is a material composed of a metal usually with a metallic material to obtain desired qualities such as improved strength, greater hardness, or resistance to corrosion.

Here is a table of some commonly used alloys.

Alloy	Composition	Characteristics	Manufacturing uses
Bronze	Copper, tin, zinc	Harder than copper, more suitable for casting, corrosion resistant	Coins, ornamental ware
Brass	Copper, zinc	Strong, ductile, corrosion resistant	Plumbing fittings
Duralumin	Aluminium, copper, magnesium, manganese	Light-weight	Aircraft frame-works
Cast iron	Iron, carbon	Improved strength	Engine blocks, machinery
Solder	Tin, lead	Melts at low temperatures	Used to join metal surfaces
Stainless steel	Iron, chromium, nickel	Corrosion resistant	Kitchen utensils
Tool steel	Iron, chromium, molybdenum	Corrosion resistant, hardwearing	Tools

DEFINITION
Degrade: reduce in strength or quality.

Finally, let's deal with resistance to degradation.

➤ Resistance to degradation

All materials degrade in time if they are allowed to be attacked by atmospheric conditions.

However, the rate of degradation varies from one material to another and can be altered by the application of protective layers, such as paint.

Let's have a look at two ways that materials are degraded: corrosion and weathering.

◆ Corrosion (or chemical attack)

Corrosion is a process in which a material is damaged by a chemical action. Metals are less resistant to degradation by corrosion than ceramics and polymers, and some metals corrode faster than others.

For example, iron and steel rust quickly under the action of oxygen and moisture. Brass, bronze and stainless steel are alloys designed to be more resistant to corrosion. Metals can be painted to slow down the corrosion process.

◆ Weathering

Weathering is a process in which a material is damaged by exposure to light, rain, or sun/heat.

Plastics may become brittle and crack in sunlight or when heated. Stabilizers such as carbon black can be added to the polymer to absorb the ultra violet light. Rubber also loses its strength in sunlight. Materials, particularly polymers, fade in sunlight. Wood needs treating to protect it from degrading by weathering.

This task is carried out in conjunction with tasks 53, 54, 55, 56, and 57

KEY CHARACTERISTICS

- physical properties
- composition
- resistance to degradation

TASK 52

Identification of the key characteristics of the materials required for the manufacture of a product

To carry out this task, use all the resources available to you such as the library, magazines, journals, teachers, tutors, local companies, and 'yellow pages'.

Identify the key characteristics of the materials required for the manufacture of each of your selected products. In task 38, you produced production plans for both of your products and these included the materials to be used in their manufacture.

Produce a list of the key characteristics of the materials required for the manufacture of your two products. One of the products must be done in detail and the other in outline.

Keep both lists in your portfolio.

4.2 PROCESSING METHODS

There is a variety of processing methods that can be used to manufacture a product. The selection of the processing method will depend upon the materials required for the product, the size and shape of the product, the number of products to be manufactured and, in most cases, the cost and time constraints that are imposed.

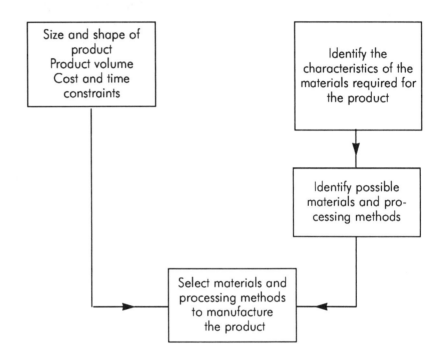

Selection of the process method

Let's look at some typical processing methods beginning with various types of heat treatment.

➤ Heat treatments

◆ Annealing

Annealing is a process of heating a material, normally glass or metal, for a given time at a given temperature, followed by slow cooling to increase ductility and strength.

◆ Moulding

Moulding is a process commonly used for shaping glass, plastics, and clays.

Injection moulding involves injecting molten plastic into a water-cooled mould which takes the shape of the mould when it solidifies. Injection moulding is used when the production volume is high and it can cater for products of widely ranging sizes and shapes.

Compression moulding involves simultaneous compressing and heating a synthetic resin powder inside a mould.

◆ Steam pressing

Steam pressing is a process used in the clothing industry where steam is used in the pressing of clothing.

◆ Pasteurization

Pasteurization is a process where milk is heated to a temperature high enough to kill any pathogenic organisms and hence make the milk safe for human consumption without spoiling its flavour and appearance.

DEFINITION
Pathogen: an organism that can cause disease

◆ Ultra-heat treatment (UHT)

Ultra-heat treatment is a process used to produce UHT milk. It uses higher temperatures than pasteurization, and kills all pathogenic organisms, giving the milk a long shelf life but altering its flavour.

◆ Canning

Canning is a process that relies on high temperatures to destroy micro-organisms and enzymes. The food is finally sealed in a can to prevent recontamination.

◆ Precipitation hardening

Precipitation hardening is a heat treatment which improves the hardness and strength of a large number of different alloys.

◆ Quenching

Quenching is a heat treatment used to harden metals. The metal is heated to a certain temperature and then quickly plunged into cold water or oil.

And now let's look at some shaping process methods.

➤ Shaping

◆ Forming

DEFINITION
Thermoplastic: plastic that always softens on repeated heating; for example, poly-ethylene and polystyrene.

Forming is a process in which thermoplastics are heated and pressed into a mould.

◆ Casting

Casting is a process in which objects are shaped by pouring a molten material of glass, plastics, or metal into a shaped mould and then allowing it to cool.

◆ Extrusion

Extrusion is a process in which materials such as plastic and metal are shaped by forcing them, usually hot, through a hole in a metal die. Tubes, sheets, and rods are commonly manufactured using this processing method.

And now let's look at some different types of surface treatment.

➤ Surface treatment

◆ Glazing

Glazing literally means to cover with a thin coat of glass.

In the food industry, glazing is applied mainly to baked pastry products to give them an attractive brown surface after baking. Glaze can be applied using a beaten egg sometimes with a little milk, or with a sprinkling of caster sugar. Tarts and flans are glazed using jelly, jam, fruit juice or syrup to give the fruit an attractive shiny finish, and prevent it from discolouring.

Ceramics such as bone china are given a vitreous coating which provides an attractive shiny finish and renders the product impervious to liquid.

Textiles are glazed using chemicals to produce a smooth, lustrous finish on a fabric.

◆ Laminating

Surfaces, particularly wooded, are covered with a thin protective layer, normally of plastic. Book jackets are normally laminated to make them look attractive and as a protection.

◆ Buffing

Metal and leather products are very often cleaned and polished using a buffer.

Let's look at some other forms of processing found in manufacturing.

➤ Other forms of processing

◆ Dyeing

Dyeing is a process of colouring textile materials by immersing them in a liquid solution of dye that provides colour resistance to washing. Dyes are either naturally occurring, such as cochineal (red) or annatto (yellow) which are both used as food colourants, or they are synthetic, such as azo dyes (red, brown, or yellow) used for printed fabrics, or mauveine (purple) used for acrylics.

◆ Mixing

Mixing, as we saw earlier in the chapter, is the combining or blending of ingredients into a single mass, ie the product. The process of mixing occurs in almost every area of manufacturing. Ingredients can be the components of a food product, or a fertilizer through to the latest drug product.

DEFINITION
Vitreous: resembling glass.

DEFINITION
Impervious: not able to be penetrated.

DEFINITION
Lustrous: having a glossy surface.

◆ Welding

Welding is a process of joining pieces of metallic material, sometimes plastic, by heat or pressure or a combination of both. Gas welding involves using the heat from a gas flame to melt the faces of the materials to be joined; arc welding uses an electric arc to perform the same action. In both cases a filler metal is used which is also melted and helps form a strong joint between the materials being welded together.

This task is carried out in conjunction with tasks 52, 54, 55, 56, and 57

TASK 53

An investigation into the processing methods used in the manufacture of a product

To carry out this investigation, use all the resources available to you.

Investigate the processing methods that would be carried out on the materials in the manufacture of your two products.

You should consider how the characteristics of the materials, and the cost and time constraints of the product determine your processing method.

Characteristics of the materials were identified and listed in task 52. Total costs of manufacturing your product were calculated in task 41. Product output was agreed in task 35.

Produce a brief report describing the processing methods used on the materials of one of your products in detail and the other in outline.

Keep this brief report in your portfolio.

4.3 HANDLING AND STORAGE OF MATERIALS AND FINISHED PRODUCTS

Storage locations in general should be:

- situated close to the working environment;
- secure;
- dry;
- well ventilated;
- well lit with light fittings placed to eliminate shadows.

They also need to:

- provide ample storage space;
- display appropriate safety notices and stickers.

The correct handling and storage of materials is very important to ensure that materials and components are available for processing in good condition.

Let's begin by looking at the handling and storage of toxic materials.

➤ Toxicity

Chemicals should be stored in a cool, dry, well-ventilated, well-drained building. There should be a procedure laid down for the way they are stored, perhaps on pallets, and there must be a safe limit on how many drums or containers are stacked on top of each other; this must be strictly enforced.

Chemicals should be stored off the ground to enable speedy identification of a leaking container. It also improves ventilation and the dispersion of flammable or toxic vapours which may result from any leakage. Inspections should be strictly carried out on a regular basis; there should also be adequate gangways to allow for this.

Chemicals should always be handled with extreme caution. A loose stopper on a drum which appears to be correctly replaced may cause an accident if the drum is bumped to the ground and chemicals are splashed on to any operatives handling the drum.

Examples of chemicals which are toxic and used in manufacturing are: lead and mercury. Mercury can be a toxic hazard even at room temperatures particularly in an ill-ventilated environment.

Trichloroethylene is used as a degreasing agent and can cause loss of control in humans; methyl chloride used as a refrigerant can cause nervous disorders if inhaled.

➤ Oxidation

Some metallic materials, especially iron and steel, undergo a chemical reaction with oxygen, particularly when moist, and form a reddish-brown oxide coating called rust. To prevent this happening or to reduce the effects of rusting, metallic materials and in particular their finished products should be stored in a dry, well-ventilated building.

FACTORS THAT AFFECT HANDLING AND STORAGE

- toxicity
- oxidation
- flammability
- perishability
- contamination
- hygiene
- sharpness
- discoloration

DEFINITION

Toxic materials: within manufacturing, chemicals used in the workplace which have a poisonous effect on the body.

FLAMMABLE MATERIALS
INCLUDE

- petrol
- thinners
- wirewool
- oils
- solvents
- paints
- gunpowder
- gases, such as butane, propane, hydrogen

Recommended temperature
ranges for various types
of storage

➤ Flammability

Flammable materials, many of which are chemicals, should in general be stored in a cool, dry, well-ventilated, well-drained building. Gas cylinders should be stored in enclosures at a safe distance from the working environment. Flammable materials must be handled with extreme care paying particular attention to naked flames and sparks which could ignite them.

➤ Perishability

Perishable materials are liable to rot and this mainly applies to food. To avoid food perishing, stock should be rotated with the oldest being used first. Food should be kept in a temperature-controlled environment to prolong its useful life.

ENVIRONMENT	TEMPERATURE RANGE
dry food stores	10°C to 15°C
cold stores	3°C to 4°C
refrigerators	-1°C to 4°C
freezers	-20°C to -18°C

➤ Contamination

Raw foods, especially meat and fish, and strong-smelling foods such as onions, cheese, and garlic should be stored separately from other food items to avoid contamination.

➤ Hygiene

Any room or area in which food is stored or handled must be clean, properly maintained and such that it can be cleaned relatively easily. Suitable lighting and sufficient means of ventilation, if appropriate, should be provided. The area should also be kept free of vermin.

Handlers of food must keep any part of their body or clothing that comes into contact with food clean. Any cuts and sores must be kept covered with a waterproof dressing. Clean and washable over-clothing must be worn. Anyone suffering from an illness which is likely to cause food poisoning must not handle food. Through the first line manager, the local medical officer is informed of the illness and the infected person is not allowed to return to handling food until the doctor gives the necessary clearance.

➤ Sharpness

Sharp materials must be treated cautiously with the correct protective clothing being worn when handling them.

Sharp materials include:

- flint;
- formica;
- bone;
- diamonds;

- glass;
- grit;
- plastic;
- any metal.

➤ Discoloration

Materials such as plastics and their finished products should not be stored outside where the weather, particularly direct sunlight, can cause fading, staining, or loss of colour.

> This task is carried out in conjunction with tasks 52, 53, 55, 56, and 57

FACTORS THAT AFFECT HANDLING AND STORAGE

- toxicity
- oxidation
- flammability
- perishability
- contamination
- hygiene
- sharpness
- discoloration

TASK 54

An investigation into the factors affecting the handling and storage of the materials used in the manufacture of a product

To carry out this investigation, use all the resources available to you.

Continue with your two products from task 53.

Investigate the factors that affect the handling and storage of the materials used in the manufacture of your two products.

Produce a summary of the factors that affect the handling and storage of the materials for each of your products.

Keep this summary in your portfolio.

4.4 PREPARATION OF MATERIALS AND COMPONENTS

The correct preparation of materials and components before the processing operations is a very important step towards the manufacture of a quality product. Sometimes it is not easy to identify what is a preparatory stage. For example, in the food industry most operations prior to processes such as baking are classed as preparatory, and in the clothing industry most operations prior to processes such as stitching are classed as preparatory.

Preparation of materials and components includes many different kinds of activity, such as:

- checking that the materials and components are not damaged;
- checking that the materials and components have been correctly stored;
- checking the size of the materials and components;
- checking the quantities of the materials and components;
- carrying out activities such as trimming, cleaning, or degreasing on the materials and components.

Let's take a look at the material and component preparations that would be carried out on a few different types of product.

➤ Cotton sports skirt

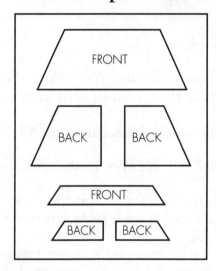

Skirt pattern

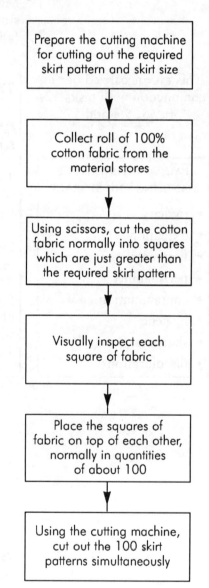

Prepare the cutting machine for cutting out the required skirt pattern and skirt size

Collect roll of 100% cotton fabric from the material stores

Using scissors, cut the cotton fabric normally into squares which are just greater than the required skirt pattern

Visually inspect each square of fabric

Place the squares of fabric on top of each other, normally in quantities of about 100

Using the cutting machine, cut out the 100 skirt patterns simultaneously

➤ Cheddar cheese

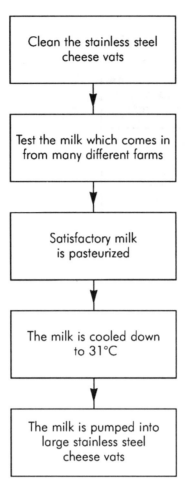

Clean the stainless steel cheese vats

↓

Test the milk which comes in from many different farms

↓

Satisfactory milk is pasteurized

↓

The milk is cooled down to 31°C

↓

The milk is pumped into large stainless steel cheese vats

➤ Metal tray

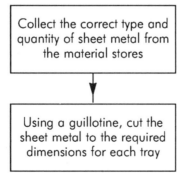

Collect the correct type and quantity of sheet metal from the material stores

↓

Using a guillotine, cut the sheet metal to the required dimensions for each tray

➤ Fish pond

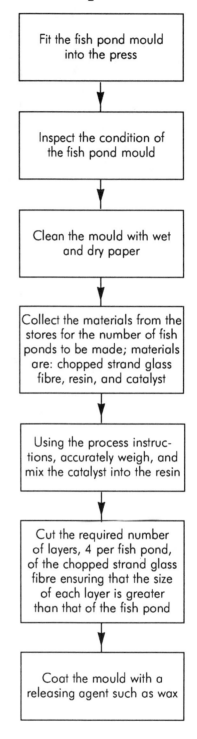

Fit the fish pond mould into the press

↓

Inspect the condition of the fish pond mould

↓

Clean the mould with wet and dry paper

↓

Collect the materials from the stores for the number of fish ponds to be made; materials are: chopped strand glass fibre, resin, and catalyst

↓

Using the process instructions, accurately weigh, and mix the catalyst into the resin

↓

Cut the required number of layers, 4 per fish pond, of the chopped strand glass fibre ensuring that the size of each layer is greater than that of the fish pond

↓

Coat the mould with a releasing agent such as wax

➤ Wooden stool

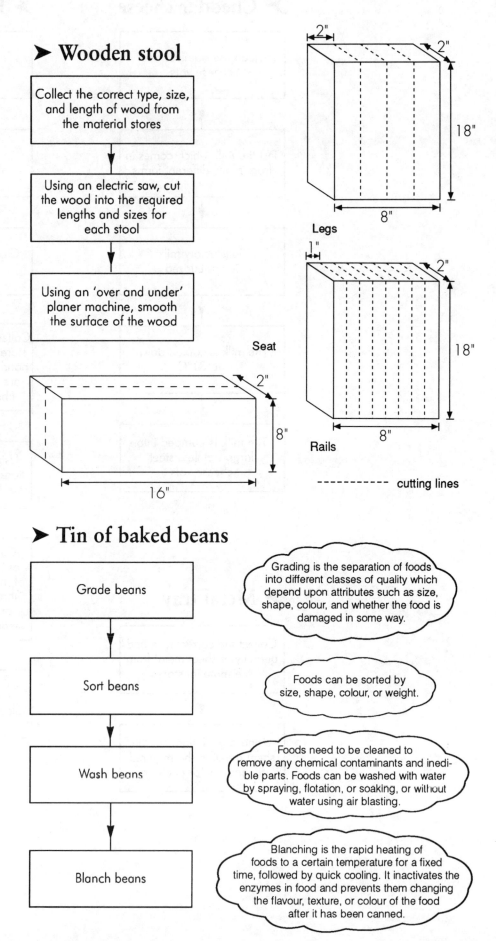

Collect the correct type, size, and length of wood from the material stores

↓

Using an electric saw, cut the wood into the required lengths and sizes for each stool

↓

Using an 'over and under' planer machine, smooth the surface of the wood

Legs
2"
2"
18"
8"

Seat
2"
8"
16"

Rails
1"
2"
18"
8"

- - - - - cutting lines

➤ Tin of baked beans

Grade beans

> Grading is the separation of foods into different classes of quality which depend upon attributes such as size, shape, colour, and whether the food is damaged in some way.

↓

Sort beans

> Foods can be sorted by size, shape, colour, or weight.

↓

Wash beans

> Foods need to be cleaned to remove any chemical contaminants and inedible parts. Foods can be washed with water by spraying, flotation, or soaking, or without water using air blasting.

↓

Blanch beans

> Blanching is the rapid heating of foods to a certain temperature for a fixed time, followed by quick cooling. It inactivates the enzymes in food and prevents them changing the flavour, texture, or colour of the food after it has been canned.

This task is carried out in conjunction with tasks 52, 53, 54, 56, and 57

TASK 55

A practical investigation into the preparation of components to specification for use in the manufacture of a product

To carry out this practical investigation, use all the resources available to you, particularly teachers and tutors.

Continue with your two products from task 54.

Prepare the materials and components to your specification listed in task 38 for use in the manufacture of at least one of your two products.

You should prepare the materials and components to meet the requirements of the production schedule which you also produced in task 38.

Produce a log briefly recording how you prepared the materials and components for use in the manufacture of at least one of your two products. If it is not possible for you to prepare the materials and components for both products, then briefly describe how you would have prepared the materials and components for the other product.

Keep the log in your portfolio.

4.5 PREPARATION OF EQUIPMENT, MACHINERY AND TOOLS

The equipment, machinery and tools must be correctly and thoroughly prepared prior to the start of manufacture. This is a very important stage in the manufacture of a quality product.

Let's begin by looking at the preparation of equipment and machinery.

Machinery can be either automated or non-automated, ie manual.

➤ Automated machinery

The controller of an automated machine is normally a minicomputer. Automated machines can be interconnected to form an automated production system. There are high investment costs, but a high output level is maintained from each machine with consistent levels of product quality.

Accurate measurements are taken automatically at each stage of the process and these are compared with the specified measurements. Automatic correction of an operation takes place if there is a trend towards the outer limits of those specified.

On the next page you will find a diagrammatic picture of automated machinery.

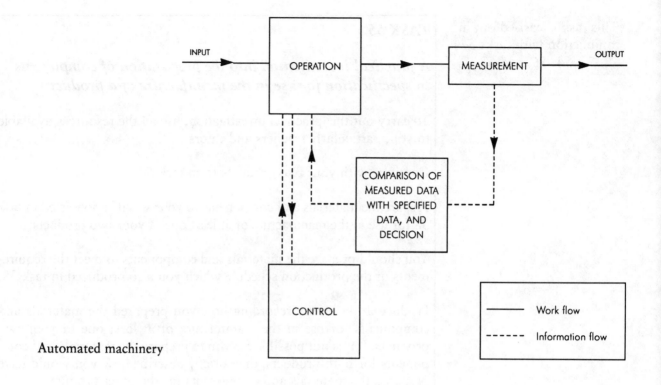

Automated machinery

➤ Non-automated machinery

A non-automated machine is one which is controlled by an operator and this can lead to variations in product quality.

The operator's vision is an integral part of the control process and, in some cases, the operator inspects and measures the processed items. Non-automated machinery can be interconnected to form a manual production system.

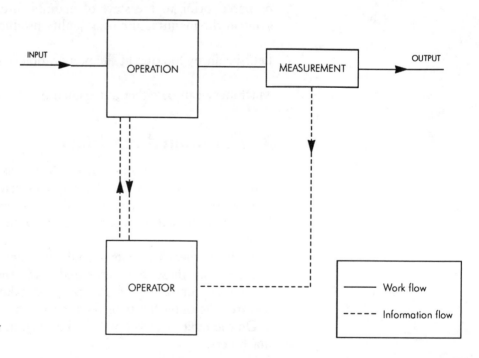

Non-automated machinery

➤ Preparation of equipment and machinery

One of the first activities in the preparation is to assess the space around the equipment and machinery.

◆ Space

There must be adequate space around a machine, particularly when the machine has an exposed blade.

◆ Floor

The floor space around equipment and machinery should be even and clean. Substances such as oil, fat or grease could cause an operative to slip and injure him/herself.

◆ Lighting

Operators must be able to see clearly all the machine controls and dials, and the item undergoing process. Lighting should be positioned so that there are no shadows over the machines.

◆ Stability

The machinery and equipment should be positioned so that the stationary parts do not move during the process operations.

The next part of the preparation is to carry out certain checks, as appropriate, on the machines and equipment.

◆ Cleanliness

A fundamental part of the preparation of machinery and equipment is to ensure that it is clean and free from dirt. This is an essential activity in the manufacturing of a quality product. Make sure that the machine is switched off at the electrical mains source before beginning to carry out any cleaning, and do *not* rely on interlock switches or machine controls. Follow the manufacturer's instructions on cleaning the machinery.

◆ Safety

Make sure that all safety devices such as guards are securely in position and serviceable.

◆ Tooling

Make sure that the correct tooling is in place for the process operation about to be carried out, if it is required.

◆ Settings

Make sure that all controls and switches are set correctly for the process about to be carried out. This will also include such actions as checking cutting edges and tensioning sewing machines.

Let's have a look at some examples of settings.

◆ Rotary oven

In a local bakery, a batch of loaves is to be baked in a rotary oven. The type of loaf being baked needs to complete one rotation (ie a revolution) every 20 seconds. What speed setting in revolutions per minute (rpm) would be needed for the baking of the loaves?

1 revolution of the loaves is completed in 20 seconds, and thus 3 revolutions of the loaves will be completed in 60 seconds, ie 3 revolutions of the loaves in 1 minute.

The rotation rate of the oven should be set to 3 rpm.

◆ Conveyor belt

In a toy factory a batch of toy bears is to be manufactured. A part of the production process requires an operator at a conveyor belt to fasten the two eyes on the toy bear securely, prior to it being 'stuffed'. The operator takes 2 seconds to remove the toy bear from the conveyor belt, 8 seconds to fasten the two eyes, and 2 seconds to replace the item on the conveyor belt. A fatigue factor of 3 seconds is allowed for the processing of each toy bear. What rate should the toy bears be fed along the conveyor belt?

Process time per toy bear is (2 + 8 + 2 + 3) seconds, ie 15 seconds. Thus the operator processes 4 toy bears in 60 seconds, ie 4 toy bears per minute.

The toy bears should be fed along the conveyor belt every 15 seconds, ie 4 toy bears per minute.

◆ Cutting machine

1. A batch of mild steel sub-assemblies requires a 10mm hole to be drilled. The recommended cutting speed of the drill is 35 metres per minute. What is the required spindle speed of the machine?

We can use the following formula to calculate SPEEDS:

$$N = \frac{1000S}{\pi d} \text{ revolutions per minute   (rpm)}$$

where:
      N is the spindle speed in rpm
      S is the cutting speed of the drill in metres per minute (m/min)
      d is the diameter of the drill in millimetres (mm)
      $\pi = 3.142$

Let's substitute the following values into the formula:
      S = 35 m/min, d = 10mm, π = 3.142

$$N = \frac{1000 \times 35}{3.142 \times 10} \text{ rpm}$$

$$N = 1114 \text{ rpm}$$

The spindle speed of machine should be set to 1114 rpm.

2. The mild steel sub-assembly is 5mm thick and the drill is required to penetrate it in 6 seconds. What is the feed rate of the drill?

We can use the following formula to calculate FEEDS:

$$F = \frac{60\,P}{Nt} \text{ millimetres per revolution } (mm/rev)$$

where:

F is the feed rate in millimetres per revolution
P is the depth of penetration in millimetres
N is the drill speed in revolutions per minute
t is the penetration time in seconds

Let's substitute the following values into the formula:
P = 5mm, N = 1114 rpm, t = 6 seconds

$$F = \frac{60 \times 5}{1114 \times 6} \text{ mm/rev}$$

$$F = \frac{300}{6684} \text{ mm/rev}$$

$$F = 0.045 \text{ mm/rev}$$

The required feed rate should be set to 0.045 mm/rev.

◆ Sewing machine

In a clothing factory a batch of waistcoats is to be manufactured. A part of the production process requires machine operators to sew together two pieces of cloth which are 0.96 metres in length, using 2mm stitches, every 2 minutes. What is the minimum stitch rate that the operators can use?

First let's calculate how many stitches are needed to sew the two pieces of cloth together.

$$\text{Number of stitches} = \frac{\text{Length of cloth}}{\text{Length of stitch}}$$

$$\text{Number of stitches} = \frac{0.96 \text{ m}}{2 \text{ mm}}$$

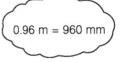

0.96 m = 960 mm

$$\text{Number of stitches} = \frac{960 \text{ mm}}{2 \text{ mm}}$$

$$\text{Number of stitches} = 480$$

Therefore 480 stitches have to be completed in two minutes,
ie 480 stitches in 120 seconds
ie 4 stitches in 1 second

The minimum stitch rate required is 4 stitches per second.

The correct preparation of equipment and machinery is vital, not just to ensure the health and safety of the operator and fellow operatives, but to minimize breakdowns during production.

➤ Preparation of tools

Finally, let's look at the preparation of tools. Tools can be broadly classified into three types: trimming, fixing, and cutting.

Let's investigate the preparation of trimming tools first.

◆ Trimming tools

A trimming tool is one used to remove surplus material around the edges of the product undergoing manufacture. Here are some examples of trimming tools used in manufacturing:

◆ Scissors

Scissors are used on textiles, paper, and food.

The correct type of scissor should be selected for the process.

Scissors should be inspected for sharpness and cleanliness

◆ Knives

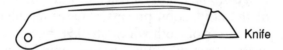

Knife

Knives are used on materials such as food, plastic, wood, and glass fibre.

The correct type of knife should be selected for the process.

Knives should be inspected for sharpness and cleanliness.

◆ Files

Files are used mainly on metals to shape and finish the surface.

The correct type and grade of file should be selected for the process.

File

Types of file include flat, half-round, and warding. Grades of file include rasp, bastard, and second-cut.

Files should be inspected for cleanliness and serviceability. Any grease or dirt should be removed using a file cleaner.

The file should not be tapped on a hard surface to remove the dirt. A handle of the correct size should be fitted over the tang of the file.

◆ Scrapers

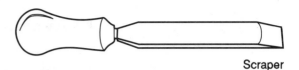

Scraper

Scrapers are used on materials such as plastic, wood, glass fibre, and metals, to remove high spots on the material surface.

The correct type of scraper should be selected for the process.

Types of scraper include, flat, half-round, and three-quarter.

Scrapers should be inspected for sharpness and cleanliness.

A handle of the correct size must be fitted to the tang of the scraper.

◆ Abrasive paper

Abrasive paper is used on wood, metal, plastic, glass fibre, etc.

The correct type and grade of abrasive paper should be selected for the process.

Types of abrasive paper include glass paper, wet and dry, and emery paper. Grades include coarse, medium, and fine.

Now let's look at fixing tools.

◆ Fixing tools

A fixing tool is used in the joining process. Here are some examples of fixing tools used in manufacturing.

◆ Sewing needles

Sewing needle

Sewing needles are used mainly on fabrics.

The correct size of needle should be selected for the process.

Needles come in ten sizes: from size 1 which is long and thick to size 10 which is short and thin.

The thickness of the fabric and thread determine the size of the needle to be used.

◆ Electric soldering irons

Electric soldering irons are used mainly with electrical and electronic components.

The correct power-rated soldering iron should be selected for the process.

Power ratings of soldering irons vary from less than one watt up to tens of watts.

The correct size soldering iron tip should be selected for the process.

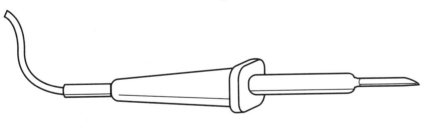

Electric soldering iron

Soldering irons should be inspected for safety and serviceability, and the tip should be clean and smooth.

◆ Screwdrivers

Screwdrivers are used mainly to turn and drive screws into materials such as wood, plastic, and metal.

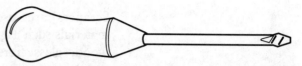
Screwdriver

The correct type and size of screwdriver should be selected for the process.

Types of screwdriver include flat blade, 'Phillips' and Pozidrive.

Screwdrivers should be inspected for damaged blades and loose handles.

◆ Spanners

Spanners are used to turn items such as nuts and bolts on materials such as metal, wood, and plastic.

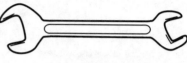
Open-ended spanner

The correct type and size of spanner should be selected for the process; using the right size is essential to prevent damage to the head of the bolt and potential injury to the operator.

Types of spanner include open-ended, adjustable, and box.

Excess oil and grease should be cleaned from the spanner.

◆ Socket spanners

Socket spanners are used to turn items such as nuts and bolts on materials such as metal, wood, and plastic.

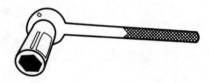
Socket spanner

The correct type and size of socket spanner should be selected for the process.

Care needs to be taken, as for spanners above.

◆ Hammers and mallets

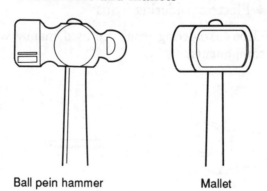
Ball pein hammer   Mallet

Hammers and mallets are used mainly with materials such as metal, wood, and plastic.

The correct type should be selected for the process.

The various types of hammer include ball pein, straight pein, and cross pein.

The various types of mallet include renewable hide or compressed fibre faces and synthetic rubber, lead, or copper.

Hammers and mallets should be inspected for loose heads and damaged handles.

And, finally, let's look at cutting tools.

◆ Cutting tools

The characteristic of a cutting tool material is that it is tough, hard, resistant to softening and generally easy to sharpen.

Here are some examples of cutting tools used in manufacturing:

◆ Scissors

Scissors are used on textiles, paper, and food.
    The correct type of scissor should be selected for the process.
    Scissors should be inspected for sharpness and cleanliness.
    In the clothing industry, small surgical-type scissors with sharp pointed ends are ideal for buttonholes and cutting off the ends of threads.

◆ Knives

Knives are used on materials such as food, plastic, wood, and glass fibre.
    The correct type of knife should be selected for the process.
    Knives should be inspected for sharpness and cleanliness.
    In the food industry it is recommended that knife blades are made of stainless steel which will not rust or stain and hence not colour food.

◆ Hacksaws

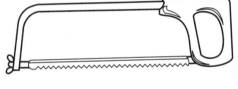

Hacksaws are used to cut off lengths of material, such as metal and glass fibre.

Hacksaw

    The correct blade should be selected for the process.
    Blades are selected according to the thickness of the material to be cut.
    A hacksaw can take blades of various lengths, the most common one being 300mm.

◆ Saws

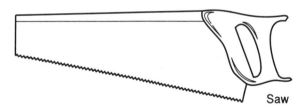

Saws are used to cut off lengths of wood.
    The correct type of saw should be selected for the process.

Saw

◆ Drills

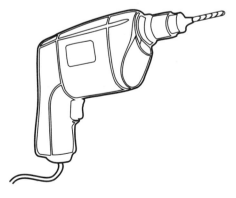

Drills are used on materials such as metal, wood, plastics, and glass fibre. They are used mainly to drill holes in those materials; electric drills also have attachments such as sanders.
    The correct drill bit size should be selected, and checked for cleanliness and sharpness.
    Electrically-powered tools should also be checked for cuts and twists in the cable, and for defective or broken mains plugs.
    Types of drill include hand drills, electric drills, and pneumatic drills.

Electric drill

◆ Guillotines

Guillotines are used to cut materials such as paper, metal, and plastics.

Before using a guillotine, the user must understand fully its safe operation, and the operation of any emergency switches or methods of immobilization. Before switching on, the user should make sure that all guards are in place and that the back of the guillotine is clear.

This task is carried out in conjunction with tasks 52, 53, 54, 55, and 57

TASK 56

A practical investigation into the preparation of equipment, machinery and tools for use in the manufacture of a product

To carry out this practical investigation, use all the resources available to you – particularly teachers and tutors.

Continue with your two products from task 55.

Prepare the equipment machinery and tools for the manufacture of at least one of your two products, to meet the requirements of the production schedule which you produced in task 38.

Produce a log briefly recording how you prepared the equipment, machinery and tools for the manufacture of at least one of your two products. If it is not possible for you to carry out this for both products, then describe how you would have prepared them for the other product.

Keep the log in your portfolio.

SAFETY EQUIPMENT, HEALTH AND SAFETY PROCEDURES AND SYSTEMS

- emergency equipment
- first aid equipment
- personal safety clothing
- safety equipment and systems
- safe working practices
- maintenance procedures
- alarms and emergency systems

4.6 SAFETY EQUIPMENT, HEALTH AND SAFETY PROCEDURES AND SYSTEMS

It is very important in a manufacturing environment that safety equipment, procedures, and systems are checked regularly and are always in place and functioning correctly whenever equipment and machinery are started up and operated.

Let's begin our investigation of safety equipment, health and safety procedures and systems by looking at emergency equipment.

➤ Emergency equipment

Emergency equipment includes fire extinguishers, hose reels, fire blankets, and sprinklers.

◆ Fire extinguishers

Fire extinguishers were traditionally painted red but nowadays under health and safety regulations they are colour coded depending on their type.

TYPE OF EXTINGUISHER	COLOUR CODE
water	red
foam	light cream
dry chemical powder	blue
carbon dioxide	black
vapourising liquid	green

Colour coding of fire extinguishers

Let's take a look at the uses of these five types of fire extinguisher.

◆ Water extinguishers

Water extinguishers are used on wood, paper, and textile fires.

The water from the extinguisher cools down the burning materials and smothers the fire with the steam that is produced.

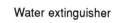

Water extinguisher

◆ Foam extinguishers

Foam extinguishers are used on fires involving flammable liquids such as oils, fats, solvents, petrol, and paint.

The foam from the extinguisher 'blankets' the fire and excludes the air.

The two main types are the plunger and the inverting foam extinguisher.

Inverting foam extinguisher

◆ Dry chemical powder extinguishers

Dry chemical powder extinguishers are used on fires involving flammable liquids both indoors and outdoors. The powder is non-toxic and can be removed easily, making it ideal for use in kitchens and food stores.

Dry chemical powder extinguisher

◆ Carbon dioxide (CO₂) extinguishers

Carbon dioxide extinguishers are used on fires involving electrical appliances and flammable liquids. The carbon dioxide gas smothers the fire.

◆ Vaporizing liquid extinguishers

Vaporizing liquid extinguishers are used on fires involving motor vehicles and electrical equipment. The vapour must reach a concentration level in excess of 9% of volume to be effective, and is therefore best used in automatic systems, for example those installed in engine bays or computer suites.

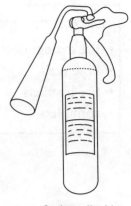

Carbon dioxide
extinguisher

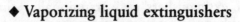

Vaporizing liquid
extinguisher

Now let's take a look at other items of emergency equipment, the next is the hose reel.

◆ Hose reels

A hose reel is not a portable item of fire fighting equipment. It is connected to the water mains and provides a powerful and sustainable jet of water to fight a small fire.

◆ Fire blankets

Fire blankets are used throughout the manufacturing industry.
They are made from synthetic fibres, commonly woven glass fibre, and are used to smother a fire.

Hose reel

Fire blanket

Finally, let's look at sprinklers.

◆ Sprinklers

Sprinklers are now installed in most factories around the UK. Sensors detect the smoke from a fire and automatically operate a sprinkler system which normally releases water on to a fire to extinguish it.

Emergency equipment should be inspected and tested regularly, once a year by law, and the findings and any remedial action carefully logged.

Now we are going to consider first aid equipment.

➤ First aid equipment

First aid equipment includes the following:

- first aid box;
- stretcher;
- blankets;
- protective clothing.

> Protective clothing may be needed if first aiders are put at risk going to the aid of an injured person.

MINIMUM CONTENTS OF A FIRST AID BOX

- copy of first aid leaflet issued by the Health and Safety Executive
- at least 6 large sterilized unmedicated dressings for general use
- at least 12 small sterilized unmedicated finger dressings
- at least 6 medium-sized sterilized unmedicated dressings for hands and feet
- at least 24 adhesive wound dressings of an approved type and in assorted sizes
- at least 4 triangular bandages of unbleached calico of the appropriate size
- a supply of adhesive plaster
- eye ointment in an approved container
- a supply of absorbent sterilized cotton in 14-gram packets
- at least 4 sterilized eye pads in sealed packets
- a rubber or pressure bandage
- safety pins

Under the Health and Safety (First Aid) Regulations 1981, the contents of first aid boxes and kits are specified, and they must be checked and replenished as necessary. An adequate number of people must be trained and hold an appropriate first aid qualification approved by the Health and Safety Executive.

Under the Approved Code of Practice issued with the Regulations, an employer should generally provide a suitably equipped and staffed first aid room where 400 or more employees are at work or in other cases where there are special hazards.

Where employees are using potentially dangerous equipment or machinery, small travelling first aid kits must be provided.

If any accident, no matter how small, occurs at work, then by law it must be entered in the accident book.

Next let's investigate the various types of personal safety clothing.

➤ Personal safety clothing

PERSONAL SAFETY CLOTHING

- approved coveralls
- safety shoes and boots
- eye protectors
- safety helmets
- respirators
- gloves and gauntlets
- ear defenders
- hats and caps

There are many different items of personal safety clothing available. Some items, such as safety shoes, are designed to protect the worker from harmful situations, for example whenever materials or equipment are lifted by the worker. Others, such as the hats worn by food industry workers, are to protect the materials from being contaminated.

Let's begin by looking at approved coveralls.

◆ Approved coveralls

These are a first line of protection, and should be close fitting and in good condition with no buttons undone or missing.

It is very important in the food industry to wear clean, washable, light-coloured protective clothing which completely covers ordinary clothing.

The coveralls are worn to protect the food from the risk of contamination by such things as woollen fibres and pet hairs. Regular checks should be carried out on the condition of the coveralls, particularly for tears and loose buttons.

Safety Clothing & Equipment

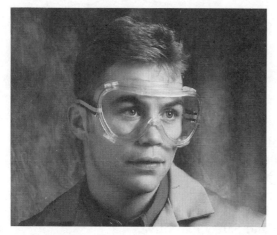

(a) Goggles

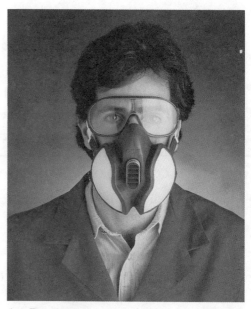

(b) Respirator

(c) Helmet with ear defender

(d) Boiler suit

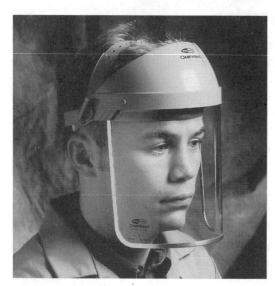

(e) Faceshield

(f) Gauntlets

(g) Boots

◆ Safety shoes and boots

The safety features include the following:

- steel toe caps;
- oil-resistant soles;
- heat-resistant soles;
- anti-static soles;
- sole plates.

Regular inspections should be carried out to make sure the shoe or boot will give adequate protection. Checks should be made for dents in the toe caps.

◆ Eye protectors

There are many different types of eye protection available. It is important to wear the correct type of eye protection for the work being carried out.

Safety spectacles with side shields provide adequate protection.

Safety goggles or full-face visors should be worn in hazardous environments – for example where there is a possibility of chemicals splashing – and should be made from special safety glass or plastic when there is a risk of projectiles.

Helmet-type screens fitted with an approved filter (dark glass) should be worn when welding, to protect the eyes from ultra violet (UV) light.

Regular checks should be carried out to ensure that the eye protection is serviceable and that no lenses are cracked.

◆ Safety helmets

Safety helmets should be worn when there is a risk from falling objects, or when entering low structures.

Regular checks should be made to make sure that the helmet is serviceable and that it has no dents.

◆ Respirators

There are several different types of respirator available, ranging from the simple dust mask which traps dust particles, to the canister respirators which absorb chemicals from the atmosphere.

Air-fed coverall suits afford full protection from dust, toxic gases and vapours.

Respirators need regular testing and checks for serviceability; canisters need to be replaced periodically.

◆ Gloves

There are several types of protective glove available.

Types of glove	Use
leather	welding
rubber	handling acids, alkalis, oils
heat-resistant	handling hot objects
plastic	food

Gloves should be worn when handling sharp objects.
Gloves need to be checked regularly for wear and tear.

◆ Ear defenders

Ear defenders range from those which fit snugly in the ear to those which cover the whole ear. They are worn for protection against noise which can affect concentration and cause hearing impairment. Ear defenders need regular checking, cleaning and disinfecting, and testing.

◆ Hats and caps

Hats and caps are worn to protect long hair from becoming entangled in machinery, or to protect materials, like food, from becoming contaminated.

The appropriate safety clothing must be used in all hazardous and hygienic environments.

Let's consider hazardous environments first.

> # Hazardous environments

HAZARDOUS ENVIRONMENTS

- hot
- cold
- contaminated
- physically dangerous
- electric shock
- hygienic

◆ Hot environment

A hot environment includes bakeries and heat treatment processes. Any type of clothing provides some protection from heat. Thick clothing made from insulation materials can temporarily prevent heat getting through to the body. Aluminized suits and aprons give greater protection against radiant heat. Gloves must be worn to protect the hands from hot materials. Pre-chilled clothing, such as ice vests, can be used to protect from heat. Air-cooled or water-cooled suits can be used but they require trailing hoses which limit mobility. In regions of intense heat, the Vortex suit can be used to protect the body.

◆ Cold environment

A cold environment includes meat and food product refrigeration. Safety clothing must provide thermal insulation and allow the evaporation of sweat. This cannot be met efficiently by a single material and in practice a triple shell material is used. A triple layer garment still needs to have adjustable ventilation to allow for variations in wind-chill and in the heat generated by work, and hence it is better to wear several removable layers than to attempt complete protection with just one garment. Terry-cloth or neoprene gloves are good for low temperature work.

◆ Contaminated environment

Examples of a contaminated environment include paint spraying and toxic fumes. Safety clothing includes respirators, full protective clothing and gloves and also the use of masks.

Respirators purify air by filtering out harmful dusts and gases, and should never be used where there is a lack of oxygen.

Breathing apparatus provides air or oxygen from an uncontaminated source via an air line or portable container.

Protective clothing should cover as much of the skin as possible to prevent any absorption of the contaminant.

A Geiger counter is an instrument used for detecting and measuring the intensity of high energy radiation.

◆ Physically dangerous environment

A physically dangerous environment includes areas containing any type of machinery, welding equipment, or lifting gear.

Various types of safety clothing are worn to protect the different parts of the body as shown in the following table:

PART OF BODY	SAFETY CLOTHING
eyes	goggles, spectacles, face screens, eye screens
hands	gloves, gauntlets
feet	safety boots, safety shoes leather spats
ears	ear plugs, muffs
head	helmet
forearms	leather sleeves
face	face guard
body	specific types of overall

◆ Electric shock environment

Most areas of work can be an electric shock environment.

Electric shock is the effect produced on the body, particularly the nervous system, when an electric current passes through it. Voltages as low as 20 volts can produce shock effects and voltages as low as 100 volts can cause death. An electric shock can cause muscles to contract. In itself this may not be serious, but it could cause injury or even death if the person is working high up, for example on a platform.

TYPES OF RESPIRATOR

CANISTER
Contains absorbent chemicals which remove gas or vapour from the air before it is breathed in. Correct canister must be used for each type of contaminant.

CARTRIDGE
Gives less protection for a shorter period than a canister respirator.

DUST
Protects against dust but not gas or vapour.

DEFINITION
Spat: another name for a gaiter, for ankle protection.

Various types of safety clothing

When working with electricity, the following safety equipment should be used to prevent electric shock:

- rubber-soled shoes/boots;
- rubber gloves;
- insulated mats and sheets;
- insulated tools and appliances.

Now let's look at the clothing that should be worn in an hygienic environment.

◆ Hygienic environment

Hygienic environments include those where food is processed. When working with food, the following should be used to protect the food from the risk of contamination:

- head covering completely enclosing the hair, such as caps and hairnets;
- clean, washable, light-coloured protective clothing, preferably without external pockets;
- ordinary clothing, worn underneath coveralls, should not protrude at all;
- thin, transparent gloves to handle food.

The next topic to investigate is safety equipment and systems.

➤ Safety equipment and systems

Let's start with guards.

◆ Guards

Guards are used to protect the operator from the dangerous moving parts of a machine.

Some guards are fixed, and their fixings must be secure and tamper-proof.

Some guards are removable in order to carry out adjustment or tool change. When this happens, the machine must be isolated, ie cut

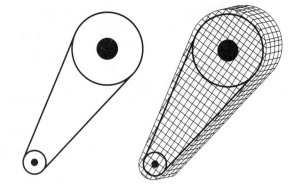

Unguarded
machine belt drive

Fixed guard
machine belt drive

off from the electrical mains supply before the guard is removed.

Guards that are removable are best fitted with interlocking switches which are connected via the electrical mains supply to the motor so that the 'dangerous part' cannot run unless the guard is in place.

Some guards are adjustable prior to machine operation, such as those

SAFETY EQUIPMENT AND SYSTEMS

- guards
- two-hand control devices
- trip devices
- warning lights and buzzers
- mechanical restraint devices
- overrun devices
- safety valves
- fuses
- circuit breakers
- residual current circuit breakers
- emergency stop buttons

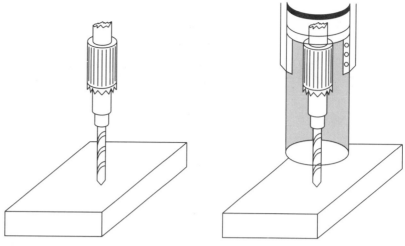

Unguarded drilling machine   Interlocking guard drilling machine

used on band saws, whilst some guards keep the 'dangerous part' out of normal reach, like those on cutting machines.

Guards must be regularly maintained. Any broken or worn parts should be repaired or replaced immediately.

◆ Two-hand control devices

A two-hand control device, as its name suggests, is a device which requires the operator to use both hands to operate the machine controls.

This device is commonly used in the shoe industry.

◆ Trip devices

A trip device is a protective appliance which activates and stops the machine once the operator (or another person) has gone beyond the safe limit of working. The activation can be through pressure sensing, ultrasonic, mechanical or photo-electric. Let's look at two examples.

Pressure sensitive mats are located at the entrances to machine bays and adjacent to hazardous machinery. Anyone stepping on the mat automatically activates a switch which causes the machine to stop.

A light curtain is a beam of light used to monitor movement. Light curtains are set up at the entrances to machine bays and adjacent to hazardous machinery; anyone breaking the beam of light automatically stops the machine.

◆ Warning lights and buzzers

Warning lights and buzzers are used to warn the operator that something needs immediate attention.

◆ Mechanical restraint devices

A mechanical restraint device is one which stops the motion of a machine when it is not working correctly. This device is commonly used on pressure die-casting machines.

◆ Overrun devices

An overrun device is used in conjunction with a guard. When a guard is removed, the power to the machine is disconnected, and during the time the machine is coming to rest, an operator may be able to access a 'dangerous part'. An overrun device prevents this access.

◆ Safety valves

A safety valve is used to prevent pressure building up in boilers, beyond their specified safety levels.

Let's now have a look at some electrical safety devices, starting with the fuse.

◆ Fuses

A fuse is used to protect electrical circuits from current overloads. It should be of the correct type and rating for the circuit or appliance it protects.

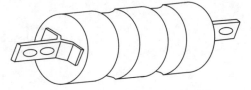

High breaking
capacity (HBC) fuse

A commonly-used industrial fuse is the high breaking capacity (HBC) fuse.

◆ Circuit breakers

A circuit breaker is a device which trips a switch from the on position to the off position if there is a current overload. The circuit breaker can be reset easily to the on position but, if possible, the reason for the current overload should be identified before doing this. The circuit breaker should be the correct type and rating for the circuit or appliance that it protects.

◆ Residual current circuit breakers

A residual current circuit breaker provides protection against earth leakage faults. To ensure that it is operating correctly, it should be tested periodically using the test button.

Finally, an extremely important safety device is the emergency stop button.

◆ Emergency stop buttons

In any manufacturing plant using electrically powered machines and equipment there should be several conveniently placed emergency stop buttons which, when pushed, will cause all electrical machinery and equipment in the plant to stop. The button is normally bright red and prominently positioned. All employees in the plant should know where these buttons are. Only an authorized person can switch the electrical supply back on again.

Machine operators should also know whether foot-operated stops or spindle brakes are fitted to their machines. These devices are used by the operative or fellow operatives to stop the machine quickly if an accident occurs.

Now let's investigate safe working practices.

➤ Safe working practices

Under the Health and Safety at Work Act, employers are responsible for ensuring the safety of their employees. This applies to the provision and maintenance of safe working conditions. Safe working practices are devised by the employer and are based on the following:

- established practices;
- makers' instructions and guidance;
- data sheets;
- Health and Safety regulations;
- established training programmes.

Supervision, training, and instructions must be given to employees, particularly when new methods and processes are introduced.

The picture shows one example of safe working practices to minimize hazards in the work place.

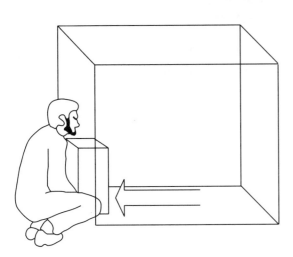

Push or pull objects in restricted areas rather than lifting them

Now we consider maintenance procedures.

➤ Maintenance procedures

Under the Health and Safety at Work Act, employers are responsible for the provision and maintenance of plant and systems of work that are safe and without risks to health. To fulfil this duty, managers must make sure that safety equipment and systems are maintained regularly and properly at all times by laying down maintenance procedures.

A manufacturing organization will have a maintenance team whose role it is to carry out these procedures; the team will include fitters, electricians, builders, and general labourers.

Finally, let's have a close look at alarms and emergency systems.

➤ Alarms and emergency systems

◆ Fire alarms

A fire alarm is used to warn all staff of the danger of a fire or an emergency. The staff should all understand the actions to be carried out when a fire alarm sounds, and should be able to evacuate the workplace safely. Fire instruction notices should be displayed at key points in the workplace, and fire drills carried out at least once each year with fire alarm sound checks being made on a regular basis.

ACTIONS TO BE CARRIED OUT WHEN FIRE ALARM IS SOUNDED

- Switch off electrical equipment
- Leave building by nearest fire exit
- Go to designated assembly point
- Do not attempt to re-enter building

This task is carried out in conjunction with tasks 52, 53, 54, 55, and 56

SAFETY EQUIPMENT

- emergency equipment
- first aid equipment
- personal safety clothing

HEALTH AND SAFETY PROCEDURES AND SYSTEMS

- safety systems
- safe working practices
- maintenance procedures
- alarms
- emergency systems

TASK 57

A practical investigation into the checking of the operation of safety equipment, and health and safety procedures and systems

To carry out this practical investigation, use all the resources available to you – particularly teachers and tutors.

Check that the safety equipment, and health and safety procedures and systems are fully operational in the manufacturing environment.

Produce a log with observations that the safety equipment and health and safety procedures are fully operational.

Bring together tasks 52–57 inclusive to produce a report titled 'Prepare materials, components, equipment, and machinery'.

Keep the report in your portfolio.

4.7 SAFE OPERATION OF EQUIPMENT, MACHINERY AND TOOLS

The safe operation of equipment, machinery and tools is very important if injuries in the workplace are to be kept to a minimum.

Let's begin by looking at the safe operation of equipment and machinery.

➤ Equipment and machinery

The safe operation of equipment and machinery in accordance with the manufacturers' instructions requires the following to be carried out:

◆ Training

Operators need to be fully trained in the operation of the equipment and machinery.

◆ Supervision

Supervisors need experience, knowledge, and time to supervise effectively; they should pay particular attention to any bad working practice which may be developing.

◆ Safety equipment and systems

Items such as guards protect the operator from the dangerous moving parts of the machine. If additional guarding is required, then it should be designed and implemented.

◆ Protective clothing and equipment

Many items of equipment and machinery cannot be operated safely without the operator wearing the correct protective clothing and equipment; if possible it should be that recommended by the manufacturer. For example, eye protection should always be worn when operating a spot welding machine.

Let's look at the safe operation of tools.

➤ Tools

We saw earlier that tools can be classified broadly into three types: trimming, fixing, and cutting.

We'll begin by looking at the safe operation of trimming tools.

◆ Trimming tools

◆ Scissors

When using scissors, the user should ensure that the free hand is kept well away from the cutting line.

◆ Knives

When using a knife, the user should direct the cut away from the body or hand. If the cut is drawn towards the body, then the item should be positioned so that should the knife slip it will pass the body or hand harmlessly.

◆ Abrasive paper

When using abrasive paper, the user should beware of hands slipping off the item and causing injury. Wrapping abrasive paper around a block of wood can provide more safety and control.

◆ Files

When using a file, the user should use both hands, one at each end, to control the file safely.

◆ Scrapers

When using a scraper, the user should direct the scraping action away from the body.

Let's consider the safe operation of fixing tools.

◆ Fixing tools

◆ Sewing needles

When using a sewing needle, the user should keep fingers well away from the sewing line.

◆ Soldering irons

When using a soldering iron, the user should hold the handle of the soldering iron firmly in one hand and carefully apply the solder using the other hand. When not in use the soldering iron should be placed in its holder.

◆ Screwdrivers

When using a screwdriver, the user should apply constant pressure to prevent it from slipping off the screw, causing injury or damage. The free hand of the user should be kept away from the screwdriver blade.

◆ Spanners

When using a spanner, the user should maintain a constant pressure on the nut or bolt. Under pressure a spanner can slip and cause injury and damage.

◆ Sockets

When using a socket, the user should maintain the turning action at a constant rate to prevent the socket slipping off the nut or bolt and causing injury and damage.

Finally, let's look at the safe operation of cutting tools.

◆ Cutting tools

◆ Scissors

When using scissors as a cutting tool, as with scissors for trimming, the user should always make sure that the free hand is kept well away from the cutting line.

◆ Knives

Just as for trimming, when using a knife, the user should direct the cut away from the body or hand. If the cut is drawn towards the body, then the item should be positioned so that should the knife slip it will pass the body or hand harmlessly.

◆ Saws

When using a saw, the user should take great care when beginning to cut. The blade should be guided carefully and steady pressure should be applied to prevent the saw from slipping. The free hand should be kept well away from the teeth of the saw.

◆ Drills

When using a drill, the work piece should be clamped firmly or held in a vice, but *not* held by hand so that if the drill bit breaks or snags (locks into the work piece) then the user is not at risk.

◆ Guillotines

When using a guillotine, the user should make sure that the guard is in place. It is dangerous to work from the back of the guillotine.

This task is carried out in conjunction with tasks 59, 60, and 61

TASK 58

A practical investigation into using the appropriate equipment, machinery and tools, safely in accordance with the manufacturers' instructions, to manufacture a product

To carry out this practical investigation, use all the resources available to you – particularly teachers and tutors.

Continue with your two products from task 57.

Use the appropriate equipment, machinery and tools, safely in accordance with manufacturers' instructions, to manufacture at least one of your two products.

Produce a log recording how you used the equipment, machinery and tools safely to manufacture at least one of your two products. If it is not possible for you to manufacture both products, then describe briefly how you would have used the appropriate equipment, machinery and tools safely to manufacture the other product.

Keep the log in your portfolio.

4.8 CONTROL AND ADJUSTMENT OF EQUIPMENT AND MACHINERY

The control and adjustment of equipment and machinery must be carried out correctly in order to process materials and components to specification and quality standards. The quality standards are specified by the customer and will include consideration of such items and materials and tolerances in order to meet the specification.

QUALITY STANDARDS

Overall quality of a product can be made up of any of the following:

Dimensions – length, diameter, thickness, etc

Physical properties – weight, volume, etc

Materials

Appearance – finish, colour, etc

Effect on senses – feel, taste, noise level, etc

Functionality – input level, output level, etc

Let's begin by looking at the control and adjustment of manual machinery.

➤ Manual equipment and machinery

A manual machine needs continuous operator attention, control and adjustment to process materials to specification and quality standards.

Some machines have to be adjusted whilst 'running', or after a trial run. The final setting of controls has to be made once the actual product has been inspected. The controls for carrying out running adjustments should be positioned safety with no necessity to remove guards.

After a trial run or inspection of the product, it may be necessary to change a particular tool or make adjustments.

Control and adjustment occurs as the manufacturing process proceeds; this can include the following:

- temperature;
- flow rate;
- voltage;
- pressure;
- speed;
- current.

EXAMPLE

Let's look at an example of pressure control and adjustment. In a canning process the food is packed into a container and then heated until the contents are sterile. A part of the heat process involves maintaining the cans at a constant pressure of 15 pounds per square inch (PSI) in a large pressure cooker for a set processing time. The pressure of 15 PSI is controlled and adjusted as necessary during the process.

Computer control systems
in paper making

➤ Automated equipment and machinery

An automated machine is pre-programmed to operate at various specific levels: for example levels of speed, temperature, feed rate, and pressure. An operator would need to observe and check that the machine is operating safely and correctly to specification whilst carrying out any necessary control and adjustments.

This task is carried out in conjunction with tasks 58, 60, and 61

TASK 59

A practical investigation into controlling and adjusting the appropriate equipment, machinery and tools, correctly in accordance with manufacturers' instructions, to manufacture a product

To carry out this practical investigation, use all the resources available to you – particularly teachers and tutors.

Continue with your two products from task 58.

Control and adjust the appropriate equipment, machinery and tools, correctly in accordance with manufacturers' instructions, to manufacture at least one of your two products to the specification listed in task 38.

Produce a log, briefly recording how you correctly controlled and adjusted the equipment, machinery and tools to manufacture at least one of your two products. If it is not possible for you to manufacture both products, then give a short description of how you would have correctly controlled and adjusted the equipment, machinery and tools to manufacture the other product.

Keep the log in your portfolio.

4.9 EXAMPLES OF PROCESSING MATERIALS AND COMPONENTS

Let's look at some typical material and component processing used in the manufacturing industry.

➤ Wooden stool

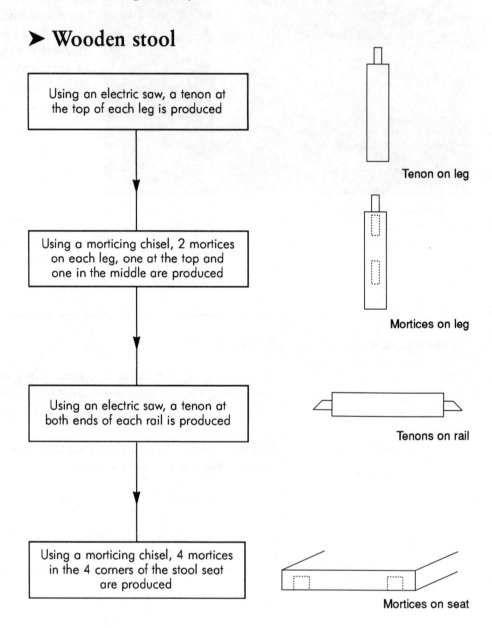

Using an electric saw, a tenon at the top of each leg is produced

Using a morticing chisel, 2 mortices on each leg, one at the top and one in the middle are produced

Using an electric saw, a tenon at both ends of each rail is produced

Using a morticing chisel, 4 mortices in the 4 corners of the stool seat are produced

Tenon on leg

Mortices on leg

Tenons on rail

Mortices on seat

➤ Cheddar cheese

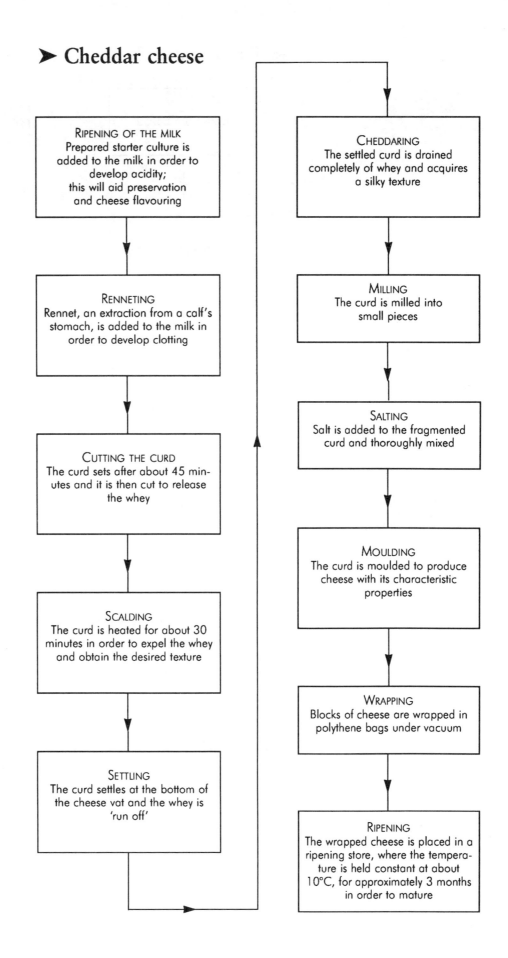

RIPENING OF THE MILK
Prepared starter culture is
added to the milk in order to
develop acidity;
this will aid preservation
and cheese flavouring

RENNETING
Rennet, an extraction from a calf's
stomach, is added to the milk in
order to develop clotting

CUTTING THE CURD
The curd sets after about 45 minutes and it is then cut to release
the whey

SCALDING
The curd is heated for about 30
minutes in order to expel the whey
and obtain the desired texture

SETTLING
The curd settles at the bottom of
the cheese vat and the whey is
'run off'

CHEDDARING
The settled curd is drained
completely of whey and acquires
a silky texture

MILLING
The curd is milled into
small pieces

SALTING
Salt is added to the fragmented
curd and thoroughly mixed

MOULDING
The curd is moulded to produce
cheese with its characteristic
properties

WRAPPING
Blocks of cheese are wrapped in
polythene bags under vacuum

RIPENING
The wrapped cheese is placed in a
ripening store, where the temperature is held constant at about
10°C, for approximately 3 months
in order to mature

➤ Tin of baked beans   **➤ Metal tray**

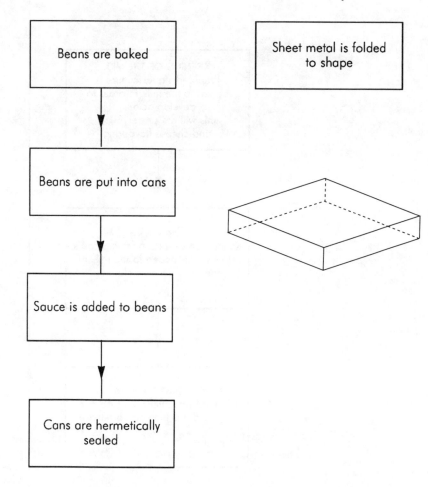

| Beans are baked |
| Beans are put into cans |
| Sauce is added to beans |
| Cans are hermetically sealed |

Sheet metal is folded to shape

Hermetically sealed: the seal is such that the object is airtight.

4.10 MAINTAINING THE LEVEL OF MATERIALS TO MEET THE PRODUCTION FLOW

We'll take an example of maintaining the level of materials required for a production system with an agreed production flow of eight products per hour.

Let's suppose the production system manufactures a product using three processes with known process times and material input levels. Look at the flow chart showing these processes within the production system on the next page.

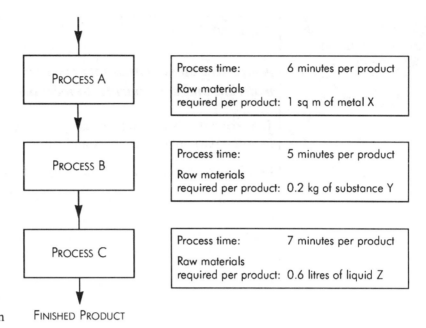

Production system   FINISHED PRODUCT

The following table shows the operation timings of the three processes over a period of one hour in the manufacture of the eight products. (All timings are inclusive.)

PROCESS A (min)	PROCESS B (min)	PROCESS C (min)	MANUFACTURED PRODUCTS
1–6	7–11	12–18	1
7–12	13–17	18–24	2
13–18	19–23	24–30	3
19–24	25–29	30–36	4
25–30	31–35	36–42	5
31–36	37–41	42–48	6
37–42	43–47	48–54	7
43–48	49–53	54–60	8

Let's now calculate the level of materials required at each process to meet the agreed production flow of four products per hour.

Process A

1 sq m of metal X is required per product
8 sq m of metal X is required for 8 products
Therefore **level of materials is 8 sq m of metal X per hour.**

Process B

0.2 kg of substance Y is required per product
1.6 kg of substance Y is required for 8 products
Therefore **level of materials is 1.6 kg of substance Y per hour.**

Process C

0.6 l of liquid Z is required per product
4.8 l of liquid Z is required for 8 products
Therefore **level of materials is 4.8 l of liquid Z per hour.**

Provided these levels of material inputs are maintained, then the agreed production flow of eight products per hour will be met.

This task is carried out in conjunction with tasks 58, 59, and 61

TASK 60

A practical investigation into correctly maintaining the levels of materials and components to meet the specified production flow

To carry out this practical investigation, use all the resources available to you – particularly teachers and tutors.

Continue with your two products from task 59.

Maintain the correct levels of materials and components required at each production stage to meet the specified production flow of at least one of your two products. Production flow can be identified from your production schedule in task 38.

Product a log, briefly recording how you maintained the correct levels of materials and components required at each production stage to meet the specified production flow of at least one of your two products. If it is not possible for you to manufacture both products, then briefly describe how you would maintain the correct levels at each stage of production of the other product.

Keep the log in your portfolio.

This task is carried out in conjunction with tasks 58, 59, and 60

TASK 61

A practical investigation into checking that safety equipment, and health and safety procedures and systems are operational when equipment, machinery and tools are used

To carry out this practical investigation, use all the resources available to you – particularly teachers and tutors.

Continue with your two products from task 60.

Check that safety equipment, and health and safety procedures are operational when equipment, machinery and tools are used during the manufacture of at least one of your two products.

Produce a log, briefly recording how you checked the safety equipment, and made sure that health and safety procedures were operational when equipment, machinery and tools were used during the manufacture of at least one of your two products. If it is not possible for you to manufacture both products, then briefly describe how you would check that safety equipment, and health and safety procedures were operational for the manufacture of the other product.

Bring together tasks 58–61 inclusive to produce a log titled: 'Process materials and components'.

Keep this log in your portfolio.

SAFETY EQUIPMENT

- emergency equipment
- first aid equipment
- personal safety clothing

HEALTH AND SAFETY PROCEDURES AND SYSTEMS

- safety systems
- safe working practices
- maintenance procedures
- alarms
- emergency systems

4.11 EQUIPMENT, MACHINERY AND TOOLS USED IN THE ASSEMBLY PROCESS

Assembly is the process of fitting or joining together the parts which make up the product.

Machinery used in the assembly process can be automated or non-automated. Industries such as food-processing involve a large amount of automation in the assembly process. However, a large number of manufacturing industries carry out the assembly process manually.

Let's begin by looking at the types of tool met in the assembly process.

TYPES OF TOOL
• trimming
• fixing
• cutting

➤ Types of tool

◆ Trimming tools

We have already met these trimming tools:

- scissors;
- knives;
- abrasive paper;
- files;
- scrapers.

Few trimming tools are used in the assembly process.

◆ Fixing tools

We have already met these fixing tools:

- sewing needles;
- soldering irons;
- screwdrivers;
- spanners;
- sockets;
- hammers and mallets.

Many of these tools are used in the assembly process.

◆ Cutting tools

We have already met these cutting tools:

- scissors;
- knives;
- hacksaws;
- saws;
- drills;
- guillotines.

Few cutting tools are used in the assembly process.

Now let's take a look at some typical equipment and machinery used in the assembly process.

➤ Equipment and machinery

◆ Sewing machine

Match the size of needle to the thickness of the fabric.
Set the type of stitch to be used.
Select the desired colour and thickness of cotton to be used.
Select the correct tension on the upper and lower thread.

◆ Power screwdriver

Select the correct type and size of screwdriver attachment.
Check that the blade of the screwdriver is not damaged.

◆ Welding equipment

There are different types of welding equipment, such as arc, gas, resistance, solvent, ultrasonic, and thermal.
You must check the equipment to make sure that it is serviceable and correctly connected.
You need to take precautions to avoid the risk of fire and explosion.
Gases such as acetylene and oxygen should be treated with care and respect.

◆ Brazing equipment

Brazing or bronze-welding uses a brass filler rod.

◆ Rivet gun

You need to fill the rivet gun with the appropriate type and size of rivet.

◆ Soldering equipment

You must select the correct wattage and size of soldering tip for the assembly process; you also need the correct size and type of solder.
You should clean the solder tip before use.

This task is carried out in conjunction with tasks 63, 64, 65, and 66

TASK 62

A practical investigation into the preparation and checking of the equipment, machinery and tools used in the assembly and finishing of a manufactured product

To carry out this practical investigation, use all the resources available to you – particularly teachers and tutors.

Continue with your two products from task 61.

Prepare and check the equipment, machinery and tools to be used in the assembly and finishing of at least one of your two selected products.

Keep a log, briefly recording how you prepared and checked the equipment, machinery and tools used in the assembly and finishing of at least one of your two selected products. If it is not possible for you to assemble and finish both products, then describe briefly how you would prepare and check the equipment, machinery and tools for the assembly and finishing of the other product.

Keep the log in your portfolio.

This task is carried out in conjunction with tasks 62, 64, 65, and 66

TASK 63

A practical investigation into the preparation and checking of materials and components used in the assembly and finishing of a manufactured product

To carry out this practical investigation, use all the resources available to you – particularly teachers and tutors.

Continue with your two products from task 62.

Prepare and check the materials and components to be used in the assembly and finishing of at least one of your two products.

Keep a log, briefly recording how you prepared and checked the materials and components used in the assembly and finishing of at least one of your two products. If it not possible for you to assemble and finish both products, then describe briefly how you would prepare and check the materials and components for the assembly and finishing of the other product.

Keep the log in your portfolio.

4.12 TYPES OF ASSEMBLY PROCESS

There are three main types of assembly process: mechanical, heat, and chemical.

Let's begin by looking at mechanical assembly processes.

➤ Mechanical processes

Mechanical fastening may be semi-permanent or permanent.

◆ Semi-permanent fastenings

These include nuts and bolts, nails, screws, locknuts, washers, and retaining rings; these are the most common method of joining materials. When dissimilar materials are to be joined together, or one side of the work is blind, then it is common to use self-tapping or drive screws. These types of screw are used to fasten anything from sheet metal and castings to plastics, fabrics, and leathers. Semi-permanent fastenings are used when maintenance is likely to be carried out.

◆ Permanent fastenings

These include rivets and metal stitching.

Rivets depend on the deformation of their structure for their holding action. Rivets are driven either hot or cold depending upon the mechanical properties of the rivet material. Aluminium rivets are cold-driven since cold working improves the strength of aluminium. Most large rivets are hot-driven.

Pop rivets are used in place of rivets in the assembly of light fabrications.

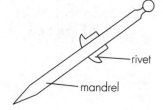

Pop rivet

Let's now consider heat assembly processes.

➤ Heat processes

Three main methods of heat fastening are welding, brazing, and soldering. Let's start with welding.

◆ Welding

Welding processes include arc, resistance, and gas to join metals; and solvent, ultrasonic, and thermal to join plastics.

◆ Arc welding

This is a process where heat is generated by an electric arc struck between a consumable welding electrode and the workpiece. Intense heat is generated by the arc and melting and subsequent solidification of the weld metal rapidly occurs. Arc welding is widely used because of its speed of operation and purity of weld. It is used in the aircraft industry and where food and drink is processed.

◆ Resistance welding

In this process an electric current flows between two thin sheet metals causing heat to be generated which joins the two metals together. Resistance welding is fast and economic and hence used in mass production systems, such as the car industry.

◆ Gas welding

Gas welding or oxy-acetylene welding is a process which uses a flame from a blowpipe to melt the edges of the parts to be joined together. Acetylene gas is stored in a maroon-coloured steel cylinder and oxygen is stored in a black drawn steel cylinder. The blowpipe is a mixing chamber for the acetylene gas and oxygen with regulating valves to vary the gas pressures.

◆ Solvent welding

This is a process used in the assembly of thermoplastic plastics where the two surfaces to be joined are placed on a pad soaked in solvent and then immediately brought together.

◆ Ultrasonic welding

In this process sonic power is applied to fuse together thermoplastics such as polystyrene. This is carried out using especially designed titanium horns. Typical products manufactured by this method are ball-point pens and camera flash cubes.

◆ Thermal welding

In this process thermoplastics are joined by first melting and then fusing together by heat. It is a popular method for joining vinyls.

Now let's look at brazing.

◆ Brazing

DEFINITION
Brazing: the process of joining metals with a non-ferrous filler metal that has a melting point below that of the metals being joined.

Brazing is the joining of similar or dissimilar metals by flowing a non-ferrous molten filler metal between them. The filler metal has a lower melting point than the metals being joined, and hence the heat required will seldom cause distortion. The melting point of the filler metal is above 800°F (427°C).

The heat for brazing is provided in many different ways, the most common being by torch, induction, furnace, and hot dipping.

BRAZING	
ADVANTAGES	DISADVANTAGES
• Dissimilar metals can be joined easily • Materials of different thicknesses can be joined easily • Joints require little or no finishing • Complex assemblies can be joined	• Limited strength in joint • Costly joint preparation • Relatively small assemblies are joined

Finally, soldering.

◆ Soldering

Soldering is a very common joining process for making electrical connections. Soldering is carried out mainly with lead-tin alloys which melt between 365°F and 450°F (ie between 185°C and 232°C).

A soldered joint is made by heating the body of the work item until the solder melts on contact and fills the joint by capillary action. Soldering is carried out by using soldering irons of different sizes and wattages, and by dips where the assembly is immersed in molten solder. Other methods of transmitting heat to the joint area are by torches, resistance heating, hot plates, ovens, induction heating, and wave soldering. A major design consideration when using soldered joints is that they should not be placed under great stress.

Now let's move on to the chemical assembly processes.

➤ Chemical processes

One of the main forms of chemical assembly processing is the use of adhesive joints, which are used extensively throughout the manufacturing industry.

Adhesives come in many different forms: solid, liquid, paste, pellet, cartridge, and tape.

Commonly used adhesives are casein, rosin, shellac, epoxies, acrylics, anerobics, urethanes, and cyanoacrylates.

What are the advantages and limitations of adhesives?

ADHESIVES	
ADVANTAGES	LIMITATIONS
• Joins practically all types of material • Results in smoother surface • Savings in weight • Savings in space • Prevents fluid leakage at joints • Prevents bimetallic corrosion at joints	• Limited working temperature • Bonding surfaces have to be very clean • Bonding is not instantaneous and requires clamping

And here are some typical uses for adhesives:

ADHESIVE	USE
Epoxy	Large joints
Acrylic	Large joints
Anerobic	Sealant
Urethane	Joints under cyclic stress

4.13 ASSEMBLY OF MATERIALS AND COMPONENTS TO SPECIFICATION AND QUALITY STANDARDS

It is important to ensure that components are assembled to the given specification and quality standards.

For many types of product there are assembly and sub-assembly drawings which are used to help achieve this. Assembly drawings, sometimes called general arrangement drawings, show what parts go together, how they are joined, and how many of each are required. When a mechanism is involved, the drawing shows how the mechanism works. Sub-assembly drawings show a part of the product and how certain components operate.

Assembly is the process of fitting parts together to create a product and this sometimes means combining the parts together to form a sub-assembly or two, and then a final assembly. To achieve this, assembly lines are used. Sometimes the assembly of the product is performed *in line*, ie along a conveyor on which the parts move, or performed *in a rotary action* with a carousel carrying the parts from position to position. The form of the assembly process must take into account the accuracy and finish of the materials and components being assembled.

4.14 EXAMPLES OF ASSEMBLY PROCESSES

Let's look at some typical assembly processes used in the manufacturing industry.

Assembling
pre-cooked meals

➤ Metal tray

SPOT-WELDING THE FOUR
CORNERS OF METAL TRAY

➤ Cotton sports skirt

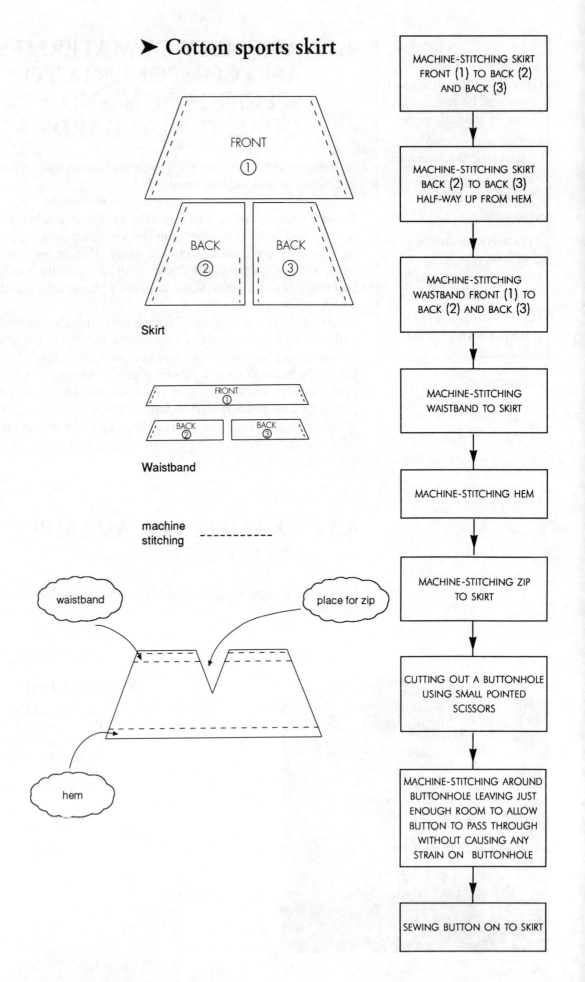

Skirt

Waistband

machine stitching ----------

waistband

place for zip

hem

MACHINE-STITCHING SKIRT FRONT (1) TO BACK (2) AND BACK (3)

MACHINE-STITCHING SKIRT BACK (2) TO BACK (3) HALF-WAY UP FROM HEM

MACHINE-STITCHING WAISTBAND FRONT (1) TO BACK (2) AND BACK (3)

MACHINE-STITCHING WAISTBAND TO SKIRT

MACHINE-STITCHING HEM

MACHINE-STITCHING ZIP TO SKIRT

CUTTING OUT A BUTTONHOLE USING SMALL POINTED SCISSORS

MACHINE-STITCHING AROUND BUTTONHOLE LEAVING JUST ENOUGH ROOM TO ALLOW BUTTON TO PASS THROUGH WITHOUT CAUSING ANY STRAIN ON BUTTONHOLE

SEWING BUTTON ON TO SKIRT

➤ Wooden stool

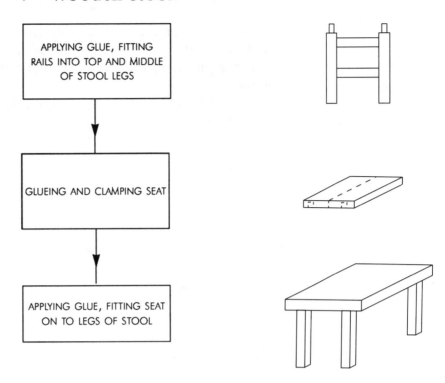

APPLYING GLUE, FITTING RAILS INTO TOP AND MIDDLE OF STOOL LEGS

↓

GLUEING AND CLAMPING SEAT

↓

APPLYING GLUE, FITTING SEAT ON TO LEGS OF STOOL

This task is carried out in conjunction with tasks 62, 63, 65, and 66

ASSEMBLY

- mechanical
- heat
- chemical

TASK 64

A practical investigation into the assembly of a product

To carry out this practical investigation, use all the resources available to you – particularly teachers and tutors.

Continue with your two products from task 63.

Prepare and check the materials, components, and sub-assemblies for the assembly process of at least one of your two products.

Assemble the materials, components, and sub-assemblies into at least one of your two products to the specification listed in task 38.

Produce a log, briefly recording how you prepared, checked and assembled the materials, components, and sub-assemblies for the assembly process of at least one of your two products.
If it is not possible for you to assemble both products, then describe briefly how you would prepare, check and assemble one of the products. (If any types of assembly, ie mechanical, heat, or chemical, are not covered in your log, then add a brief description of them along with typical products assembled by that method.)

Keep the log in your portfolio.

4.15 FINISHING METHODS

Finishing is an important stage in the manufacture of a product, particularly a quality one. Finishing methods vary from gilding ornaments to polishing furniture, and from labelling a bottle of wine to ironing a dress.

There are different reasons for applying a finish to a product and these include:

- decoration;
- protection of surface;
- corrosion-resistance;
- providing a hard surface.

The finishes applied to a product should have qualities such as:

- non-fading colours;
- uniform covering;
- freedom from runs, checks, or peelings;
- ability to expand or contract under weather or operating conditions;
- hard surface that can be cleaned and will not allow dust or grit to embed, or oil to stain.

A product may have one of a range of different finishes applied to it; for example a cookware product may be uncoated or coated in a heat-resistant plastic such as Teflon.

Products may have more than one finish applied to them; for example upholstery may be both stain-resistant and flame-resistant.

Let's take a look at some typical finishes used in the manufacturing industry.

◆ **Fabric finishes**

- crease-resistance
- waterproofing
- stain-resistance
- water-resistance
- colourfastness
- drip-dryability
- washability
- durable press
- durable pleats and creases
- flame-resistance

◆ **Plastic finishes**

- spray painting
- silk-screen printing
- vacuum metallizing
- electroplating
- roller coating
- flash removal
- buffing
- polishing
- ashing

◆ **Wood finishes**

- varnish
- paint
- sealant
- polish

◆ **Food finishes**

- icing sugar
- coarse sugar
- bun wash
- glacé cherries, almonds, desiccated coconut
- cinnamon sugar

FINISHING METHODS

- applied liquids
- coatings
- heat treatments

There is a vast number of finishing methods which can be classified broadly into three types.

Let's investigate applied liquids first.

➤ Applied liquids

First we'll look at two important types called organic and inorganic finishes.

◆ Organic finishes

DEFINITION
Organic: of, relating to, or derived from living plants and animals.

Organic finishes usually consist of two coats: firstly a priming coat and secondly a top coat. An organic coating forms a decorative and protective barrier on the base metal to which it is applied.
Organic finishes are applied by:

- roller coating;
- tumbling;
- flow coating;
- silk screen;
- centrifugal process.
- spraying;
- brushing;
- electrocoating;
- dipping;

◆ Inorganic finishes

Inorganic finishes are harder, more rigid and more resistant to increased temperatures than organic finishes.
Their main characteristics are:

- corrosion resistance;
- thermal insulation;
- eye-appealing finish;
- oxidation resistance.

There are two inorganic coatings that are widely used in the manufacturing industry: porcelain enamels and ceramic coatings.

We'll begin by looking at porcelain enamels.

◆ Porcelain enamels

These are:

- easily cleaned;
- colour stable.
- durable;

The biggest users of porcelain enamels are manufacturers of:

- kitchen ranges;
- refrigerators;
- heating equipment;
- water heaters.
- washing machines;
- dishwashers;
- dryers;

◆ Ceramic coatings

DEFINITION
Vitreous: of, or resembling glass.

They are a vitreous coating of a hard, glass-like material. They are used in applications where it is required to protect metal surfaces from the destructive effects of corrosion at high temperatures.

Ceramic coatings are used in:

- jet aircraft engines;
- chemical industry;
- textile industry;
- space vehicles;
- steel industry;
- data-processing industry.

Let's now investigate coatings.

➤ Coatings

COATINGS
- metallic
- conversion

There are many different types of coating used on products. Two important types are metallic and conversion.

◆ Metallic coatings

A metallic coating is the deposition of metal on a base metal or non-metal.

Metallic coatings are applied for:

- protection against corrosion;
- decorative finish;
- wear resistance;
- increased dimensions of a part.

Metallic coatings serve as a base for painting, to provide a:

- reflective surface;
- electrical conductive surface.
- thermal conductive surface;

Metallic coatings can be applied by these methods:

- electroplating;
- dipping;
- diffusion;
- metallizing;
- vapour;
- galvanizing.

The main users of electroplating are:

- automobile industry;
- radio manufacturers;
- television manufacturers;
- boat builders.

Metallizing can be applied to metal, plastics, wood, paper, and leather. It is used on brake discs, clutch pads, and electrical contacts.

Next let's look at conversion coatings.

◆ Conversion coatings

Conversion coatings are produced when a film is deposited on the base metal as a result of chemical reaction.

Main conversion types are:

- chromate coatings;
- anodized coatings;
- phosphate coatings;
- oxide coatings.

◆ Chromate coatings

- applied by immersion, spray, brush, roller
- thin coating
- provide high corrosion protection
- used on non-ferrous materials

◆ Phosphate coatings

- applied by immersion or spraying
- excellent resistance to humidity and weathering
- green coloured
- used in refrigerators, freezers, air conditioners, automobiles, metal furniture

◆ Anodized coatings

- electrochemical treatment of aluminium and magnesium
- clear or coloured coatings
- used on aircraft and automobile components, electronic components, furniture, sporting goods, jewellery, food industry equipment

◆ Oxide coatings

- mainly decorative
- provide corrosion resistance
- increase wear resistance

Other types of coating include polishing and waxing.

Finally, let's consider the various types of heat treatment.

HEAT TREATMENTS
• annealing
• normalizing
• hardening

➤ Heat treatments

Let's begin with annealing.

◆ Annealing

Annealing is the heating of a material, generally plastics and metal, to a certain critical temperature, and then slowly cooling it back to room temperature at a controlled, predetermined rate. In the textile industry, thermoplastic fibres such as polyester are heated to high temperatures and then cooled, after which the fabric retains its shape; for example, forming permanent pleats in garments.

◆ Normalizing

Normalizing is similar to annealing, but the material is air cooled.

◆ Hardening

Principal methods of hardening metals are cold working, precipitation, and quench hardening.

◆ Cold working

This is also known as mechanical working, and it increases the hardness or strength of a metal.

◆ Precipitation

Precipitation is also known as age hardening, and is used to improve the mechanical properties of a metal.

◆ Quench hardening

This is also known as harden and temper and is the usual method of hardening a piece of steel.

Bread and pastry products are hardened by baking. Bricks are hardened in a kiln. Porcelain is hardened by low temperature firing between 900 and 1000°C, making it strong enough to withstand being dipped in a glaze slip. Sanitary ware is produced by a firing process in a kiln at a temperature between 1150 and 1250°C. Integrated circuit boards are fire glazed.

4.16 EXAMPLES OF FINISHING PROCESSES

Let's look at some typical finishing processes used in the manufacturing industry.

➤ Cotton sports skirt

```
┌─────────────────────────────┐
│  SEWING 3 LABELS INTO SKIRT: │
│      I) 100% COTTON,         │
│  II) BRAND NAME, III) SIZE   │
└─────────────────────────────┘
              │
              ▼
┌─────────────────────────────┐
│  CUTTING OFF EXCESS COTTON   │
│     THREADS FROM SKIRT       │
└─────────────────────────────┘
              │
              ▼
┌─────────────────────────────┐
│        IRONING SKIRT         │
└─────────────────────────────┘
```

➤ Metal tray

```
┌─────────────────────────────┐
│    GALVANIZING METAL TRAY    │
└─────────────────────────────┘
```

➤ Wooden stool

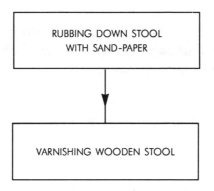

➤ Cheddar cheese

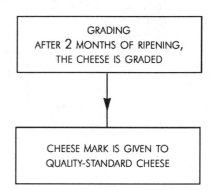

This task is carried out in conjunction with tasks 62, 63, 64, and 66

FINISH
- applied liquids
- coatings
- heat treatments

TASK 65

A practical investigation into the finishing of a product

To carry out this practical investigation, use all the resources available to you – particularly teachers and tutors.

Continue with your two products from task 64.

Finish at least one of your two products to meet the specification listed in task 38.

Produce a log, briefly recording how you finished off at least one of your two products. If it is not possible to finish off both products, then describe briefly how you would finish off one of the products.

(If any finishing methods, ie applied liquids, coatings, or heat treatments, are not covered so far in your log, then add a brief description of them along with typical products finished by that method.)

Keep the log in your portfolio.

This task is carried out in conjunction with tasks 62, 63, 64, and 65

SAFETY EQUIPMENT

- emergency equipment
- first aid equipment
- personal safety clothing

HEALTH AND SAFETY PROCEDURES AND SYSTEMS

- safety systems
- safe working practices
- maintenance procedures
- alarms
- emergency systems

TASK 66

A practical investigation into using safety equipment, and health and safety procedures and systems correctly during assembly and finishing

To carry out this practical investigation, use all the resources available to you – particularly teachers and tutors.

Continue with your two products from task 65.

Use the safety equipment, and health and safety procedures and systems correctly during the assembly and finishing of at least one of your two products.

Produce a log, briefly recording how you used the safety equipment, and health and safety procedures correctly during the assembly and finishing of at least one of your two products. If it is not possible for you to manufacture both products, then describe briefly how you would check that safety equipment, and health and safety procedures were used correctly for the assembly and finishing of one of the products.

Bring together tasks 62–66 inclusive to produce a log titled: 'Assemble and finish materials and components to specification'.

Keep the log in your portfolio.

EXAMPLES OF QUALITY INDICATORS

- weight
- volume
- size
- functionality
- finish
- taste
- sound
- smell
- touch
- appearance

4.17 QUALITY INDICATORS AND FREQUENCY OF ANALYSIS

➤ Quality indicators

We met quality indicators in section 3.17, and saw that a quality indicator is a variable or an attribute of a product that can be respectively measured or assessed. The data obtained from measurement or assessment can be compared with the product's specification to give an indication of its quality.

We also saw in section 3.17, two examples of the identification of quality indicators for the product specifications of a metal tray and a sports skirt. Quality indicators are applied at critical control points which are identified at different stages of production.

➤ Frequency of analysis

Generally, if the volume of products manufactured is small then every product is inspected, ie 100 per cent inspection. However, when there are large volumes of products being manufactured, it is not economic (it is time-consuming and costly) to inspect every product manually. In many continuous production systems the inspections are automated and 100 per cent inspections are carried out. In batch production systems, samples are taken from each batch and a decision about the quality of a batch of products is made after analysing the sample.

We met sampling in section 3.15 and found that if the sample of products conforms to the specified quality levels, then the whole batch is accepted, or if it does not conform, then the whole batch is either rejected or subjected to further inspection.

The sample size that we inspect must be large enough to indicate the true quality condition of the whole batch. If it is too small, then we may get a false indication of the quality of the products; if it is too large, then it will probably result in over-inspection (uneconomic).

The sample size may be a portion of a whole batch or the number inspected every hour.

The following two examples show the frequency of analysis and quality indicators at critical control points in the manufacture of a metal tray and a sports skirt.

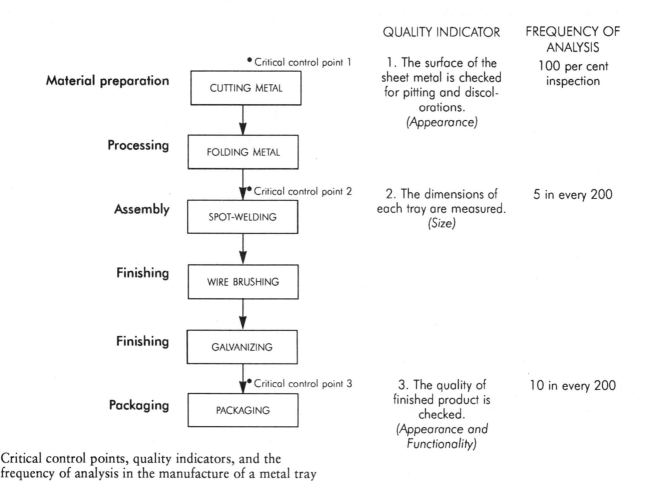

		QUALITY INDICATOR	FREQUENCY OF ANALYSIS
Material preparation	• Critical control point 1 CUTTING METAL	1. The surface of the sheet metal is checked for pitting and discolorations. (Appearance)	100 per cent inspection
Processing	FOLDING METAL		
Assembly	• Critical control point 2 SPOT-WELDING	2. The dimensions of each tray are measured. (Size)	5 in every 200
Finishing	WIRE BRUSHING		
Finishing	GALVANIZING		
Packaging	• Critical control point 3 PACKAGING	3. The quality of finished product is checked. (Appearance and Functionality)	10 in every 200

Critical control points, quality indicators, and the frequency of analysis in the manufacture of a metal tray

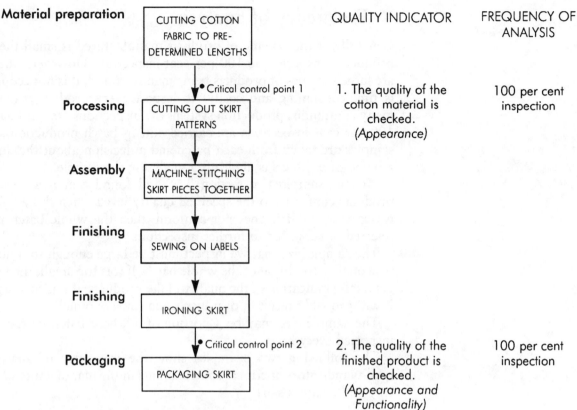

Material preparation	CUTTING COTTON FABRIC TO PRE-DETERMINED LENGTHS	QUALITY INDICATOR	FREQUENCY OF ANALYSIS
Processing	• Critical control point 1 CUTTING OUT SKIRT PATTERNS	1. The quality of the cotton material is checked. (Appearance)	100 per cent inspection
Assembly	MACHINE-STITCHING SKIRT PIECES TOGETHER		
Finishing	SEWING ON LABELS		
Finishing	IRONING SKIRT		
Packaging	• Critical control point 2 PACKAGING SKIRT	2. The quality of the finished product is checked. (Appearance and Functionality)	100 per cent inspection

Critical control points, quality indicators, and the frequency of analysis in the manufacture of a sports skirt

---

This task is carried out in conjunction with tasks 68, 69, 70 and 71

SCALE OF PRODUCTION

- continuous flow or line
- repetitive batch
- small batch or single item

TASK 67

An investigation into identifying quality indicators and the frequency of analysis for a product specification

To carry out this investigation, use all the resources available to you – such as teachers, tutors, the library, magazines, journals, local companies, and 'yellow pages'.

Continue with your two products from task 66.

In task 51, you applied quality control (quality indicators and critical control points) to your two products. You produced a flow diagram identifying the critical control points and the quality indicators to be used at these points for each product.

Reproduce the flow diagram for each of your two products, showing the key stages of production and the quality indicators

Choose a different scale of production for the manufacture of each product.

Estimate the frequency of analysis for every quality indicator for each of your two products at the different scales of production and put them on your flow diagrams.

Keep the flow diagrams in your portfolio.

4.18 TEST AND COMPARE SAMPLES AT CRITICAL CONTROL POINTS

In section 3.14 we met the role of testing and comparison in quality control, which is carried out to assess whether a product conforms to its specification. We also saw, in section 3.15, that there are various inspection and testing methods used within quality control.

Visual methods include inspecting the colour, finish, or texture of the manufactured product.

Mechanical methods are generally classified as destructive and non-destructive testing. The former includes tensile, hardness, and fatigue testing, whilst the latter includes X-ray inspection and testing by micrometer.

Electronic and electrical inspection and testing methods include functionality testing of a product.

Expert scrutiny is used for the inspection and testing of musical instruments and banknotes among other things.

Chemical analysis includes the inspection and testing of products for their composition and pH balance.

The following two examples show the testing and comparison of samples at critical control points in the manufacture of a metal tray and a sports skirt.

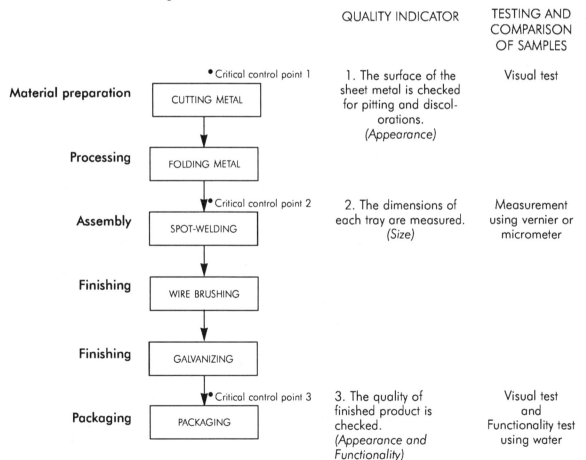

Critical control points, quality indicators, and testing and comparison of samples in the manufacture of a metal tray

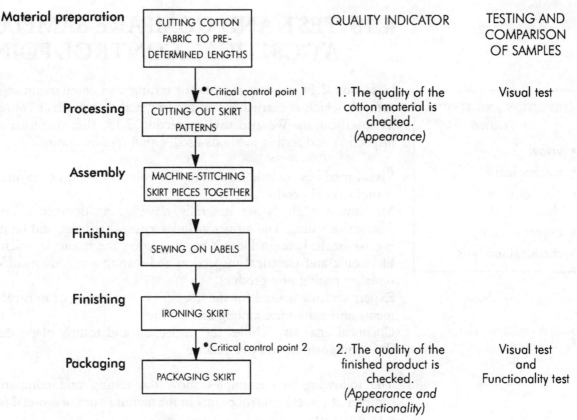

Material preparation		QUALITY INDICATOR	TESTING AND COMPARISON OF SAMPLES
	CUTTING COTTON FABRIC TO PRE-DETERMINED LENGTHS		
Processing	• Critical control point 1 CUTTING OUT SKIRT PATTERNS	1. The quality of the cotton material is checked. *(Appearance)*	Visual test
Assembly	MACHINE-STITCHING SKIRT PIECES TOGETHER		
Finishing	SEWING ON LABELS		
Finishing	IRONING SKIRT		
Packaging	• Critical control point 2 PACKAGING SKIRT	2. The quality of the finished product is checked. *(Appearance and Functionality)*	Visual test and Functionality test

Critical control points, quality indicators, and testing and comparison of samples in the manufacture of a sports skirt

---

This task is carried out in conjunction with tasks 67, 69, 70 and 71

TASK 68

An investigation into testing and comparing samples at the critical control points of the production process

To carry out this investigation, use all the resources available to you.

Continue with your two products from task 67.

In task 51, you applied quality control (quality indicators and critical control points) to your two products. You produced a flow diagram identifying the critical control points and the quality indicators to be used at these points for each product.

Reproduce the flow diagram for each of your two products, showing the key stages of production and the critical control points. Using the different scales of production that you chose in task 51, determine the method of testing and comparing to be used at each critical control point, and put them on your flow diagrams.

Produce a flow diagram for each product, showing the key stages of production, critical control points and method of testing and comparing. Use the same scales of production that you chose in task 51.

Keep the flow diagrams in your portfolio.

4.19 FORMATS FOR RECORDING DATA

Data can be recorded manually or using a computer, and in graphical, tabular, or written format.

A computer is useful for analysing large amounts of data and presenting them in different formats, whereas a manual data recording system can only be effective used on small quantities of data where simple graphs and tables are required.

Computers are used extensively in quality control for the collection, analysis, and reporting of data. Data is collected from such sources as process inspection areas, defective products, scrap and re-workings, customer complaints, and process control. The data is analysed and reports are produced normally on a weekly or monthly basis. The computer can be programmed to notify the user immediately of any potential quality problems. This allows the user to take the necessary corrective action prior to the issue of the report.

Computers are also used in the areas of process control for automatically measuring, analysing and controlling process variables such as voltage, temperature, pressure, and weight. Automated process control systems are used for example in the drink and paper-making industries.

Finally, a very important use of computers in quality control is for statistical analysis which eliminates time-consuming manual calculations and allows the data to be presented in graphical and tabular form.

Let's take a look at a simple example of testing and comparing a sample of a batch of products at a critical control point during their manufacture, and recording the data collected in an appropriate format.

Suppose a baker produces 1 000 bread rolls from pieces of dough automatically cut to weigh 39g. A sample of 40 pieces of dough is taken for testing, using the quality indicator of weight.

This table shows the results of accurately weighing the 40 pieces of dough to the nearest gram.

Test piece	Weight (g)	Test piece	Weight (g)	Test piece	Weight (g)	Test piece	Weight (g)
1	36	11	39	21	38	31	39
2	36	12	38	22	38	32	39
3	37	13	42	23	40	33	37
4	38	14	41	24	39	34	39
5	37	15	40	25	39	35	41
6	40	16	40	26	37	36	39
7	41	17	39	27	40	37	40
8	41	18	40	28	39	38	38
9	40	19	39	29	42	49	39
10	39	20	39	30	38	40	39

Table of results of weighing a sample of 40 pieces of dough

The collected data is not sorted and is difficult to analyse. We can use the computer to organize it for us. Let's present the data in the form of a frequency distribution table with the weights of the dough pieces in rank order, ie either start with the smallest weight value and work through to the largest value or vice versa, and at each weight value state its frequency of occurrence.

Weight (g)	Frequency
36	2
37	4
38	6
39	14
40	8
41	4
42	2
Total number	40

Frequency distribution of the results of weighing a sample of 40 pieces of dough

The frequency distribution provides a reasonable presentation of the data collected from the sample. Fourteen pieces of dough are of the correct weight and the remainder are spread more or less evenly either side of 39 grams.

EXAMPLE

If, for example, the weight of the dough pieces is acceptable between 39 ± 1g, then 28 dough pieces are acceptable and 12 dough pieces are unacceptable.

This means that $\frac{12}{40}$, ie 30% are unacceptable.

From this frequency distribution, a histogram can be drawn which provides a graphical representation of the weight of the dough pieces in the sample.

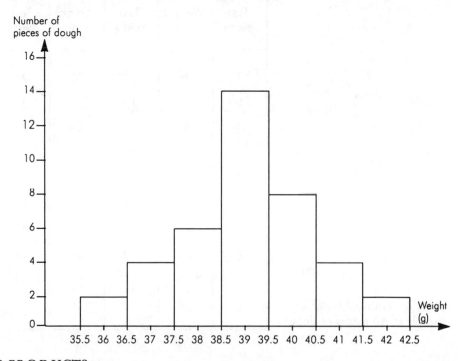

Histogram of the results of weighing a sample of 40 pieces of dough

The histogram provides another useful presentation of the data collected from the sample.

The data collected from weighing the dough pieces is continuous, ie can take any value in a given range, as compared with being discrete which takes limited values. This means that a piece of dough weighing between 39.5g and 40.5g is noted as 40g.

Weights of 39.5g to 40.5g is called a class with 39.5g the lower class limit and 40.5g the upper class limit, and the difference of 40.5 – 39.5, ie 1.0 being the class width. The frequency distribution can be presented as follows:

Weight (g)	Frequency
35.5 – 36.5	2
36.5 – 37.5	4
37.5 – 38.5	6
38.5 – 39.5	14
39.5 – 40.5	8
40.5 – 41.5	4
41.5 – 42.5	2
Total number	40

Frequency distribution of the results of weighing a sample of 40 pieces of dough

Another useful method for presenting data is the cumulative frequency table. The cumulative frequency of any class is the total frequency up to and including the upper limit in that class.

Let's construct a cumulative frequency table for the frequency distribution of the pieces of dough, using the upper class limit of each class.

Weight (g)	Frequency	Cumulative frequency
36.5	2	2
37.5	4	4 + 2 = 6
38.5	6	6 + 6 = 12
39.5	14	14 + 12 = 26
40.5	8	8 + 26 = 34
41.5	4	4 + 34 = 38
42.5	2	2 + 38 = 40

Cumulative frequency distribution of the results of weighing a sample of 40 pieces of dough

Now let's draw a cumulative frequency curve called an ogive (pronounced 'oh-jive').

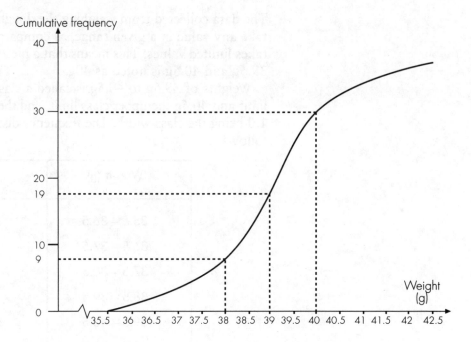

Cumulative frequency curve of the results of weighing a sample of 40 pieces of dough

The cumulative frequency curve shows us that between the acceptance limits of 38 and 40 there are (30 – 9), ie 21 dough pieces. There are nine below the acceptance limits and ten above the acceptance limits.

This task is carried out in conjunction with tasks 67, 68, 70 and 71

TASK 69

Recording data from testing and comparing a sample and displaying it in an appropriate format

To carry out this activity, use all the resources available to you.

Continue with your two products from task 68.

In task 68, you determined the method of testing and comparing at each critical control point for each of your products at different scales of production.

For at least one of the critical control points of each product, produce a table of results which would typically represent the testing and comparison of a sample of the product. From your table of results, produce the following:

- table of frequency distribution
- histogram
- table of cumulative frequencies

If it is possible, produce the above tables and graphs manually for one of your products, and using a statistical software package for the other.

Keep both sets of results, tables, and graphs in your portfolio.

4.20 DEFECTS AND THEIR CAUSES

A defect is a fault or flaw in a product which does not comply with the required specification.

➤ Types of defect

Four important types are:

- critical defect;
- major defect;
- minor defect;
- incidental defect.

Let's consider the critical defect first.

◆ Critical defect

A critical defect is one likely to result in hazardous or unsafe conditions for people who use or come into contact with the product.
Example: a defect in a car braking system.

◆ Major defect

A major defect is one likely to result in failure or reduction in the use of the product.
Example: a blocked nozzle on a spray aerosol.

◆ Minor defect

A minor defect is one likely to have little or no effect on the use of the product.
Example: a loose thread on a cotton towel.

◆ Incidental defect

An incidental defect is one that has no effect on the use of the product.
Example: a knot occurring in wood.

➤ Causes of defect

POSSIBLE CAUSES OF
DEFECT

- operator errors
- change of conditions in a process
- change of proportions in a composition
- substitution of materials

There are many possible causes of defects in the manufacture. Let's take a look at some of the more common ones, starting with operator errors.

◆ Operator errors

Operator errors normally occur in two ways: by inadvertent errors and technique errors. An inadvertent error, as its name suggests, occurs at any time and is due primarily to lack of attention by the operator. To improve the situation, processes are 'fool-proofed', so that the extent of dependence on human attention is reduced. Job rotation and rest periods are used to make it easier for operators to remain attentive. Technique errors occur when an operator lacks a particular skill or knowledge. This problem is resolved by providing the operator with extra training.

◆ Change of conditions in a process

The conditions within a process can change because of poor maintenance or faulty components.

Example: a beverage firm uses an automatic process to fill jars with 100g of coffee. If conditions change and the process underfills the jars such that the weight of the coffee is outside the acceptable limits, then the quality of the product is reduced.

Another example is perhaps that the temperature setting on an oven in a bakery is registering a lower reading than the actual temperature. Products are therefore being slightly over-baked and quality is being reduced.

These types of problem need to be rectified immediately and regularly checked during maintenance.

EXAMPLES

◆ Change of proportions in a composition

A change of proportions in the composition of a product is a common cause of reduced quality.

EXAMPLE

Example: a jam maker manufactures 1lb (454g) jars of jam and two of the major ingredients are fruit and sugar, 23g of fruit and 6.5g of sugar per 100g of jam. If the composition of the fruit and sugar fed into the mixing process changes from the stated specification, then there is a possible cause of reduced quality.

This type of problem needs to be rectified immediately and regularly checked during maintenance.

◆ Substitution of materials

The substitution of materials in a product can have an effect on product safety, and product designers need to carry out stringent tests before this is allowed to occur.

The substitution of materials is normally due to costs, and very often sub-standard materials are used, having an effect on the quality of the product.

This task is carried out in conjunction with tasks 67, 68, 69, and 71

TASK 70

An investigation into identifying defects and their possible causes and preventions

To carry out this investigation, use all the resources available to you – such as teachers, tutors, the library, magazines, journals, local companies, and 'yellow pages'.

Continue with your two products from task 69.

For each of your two products identify at least three possible defects, and for each defect suggest a possible cause and method of preventing it.

Produce a brief *written* report for both of your products, identifying at least three possible defects with possible causes and methods of prevention.

Keep the reports in your portfolio.

4.21 DEALING WITH DEFECTIVE PRODUCTS

A defective product is one that is unacceptable because it does not comply with the product specification. For example, it has a poor appearance. The action carried out on a defective product depends upon the extent or the severity of the defect.

➤ Procedures

◆ Investigate cause

A product with a quality deficiency should be identified by the quality control systems through inspections at the critical control points. If control charts are involved, then these will be passed to a quality engineer for analysis. Once the defect is identified, a person, such as a planning engineer, is made responsible for taking corrective action to remove the cause of the defect.

The cause of the defect is likely to be due to any of the following:

- unsatisfactory tools and equipment;
- unsatisfactory materials;
- ill-trained operators;
- unsatisfactory production methods;
- unsatisfactory quality control system;
- unsatisfactory product design.

◆ Recommend solutions

The solutions recommended to remove the cause of a product defect are likely to come from one of the following:

- improved tools and equipment;
- alternative or better quality materials;
- extra or improved training of operators;
- improved production methods;
- efficient quality control system;
- improved product design.

◆ Re-work

Re-working is the correction of a defective product to meet the required product specification.

◆ Scrap

Scrap is a defective product which cannot be re-worked, used, or sold, but it can be recycled.

◆ Downgrade

Downgrading occurs if a defective product is usable but does not meet the required product specification, and is sold as 'seconds quality' at a reduced price.

PROCEDURES FOR DEALING WITH DEFECTIVE PRODUCTS

- investigate cause
- recommend solutions
- re-work
- scrap
- downgrade

DEFINITION
Cause: a proved reason for the existence of the defect.

This task is carried out in conjunction with tasks 67, 68, 69, and 70

TASK 71

An investigation into the procedures for dealing with defective products

To carry out this investigation, use all the resources available to you – such as the library, magazines, journals, teachers, tutors, local companies, and 'yellow pages'.

Identify the procedures that should be used when dealing with defective products.

Produce a brief *written* report outlining your suggested procedures when dealing with defective products.

Bring together tasks 67–71 inclusive to produce a report titled: 'Apply quality assurance to manufactured products'.

Keep the report in your portfolio.

PROCEDURES

• investigate cause
• recommend solutions
• re-work
• scrap
• downgrade

This is a stand-alone task and will not form part of any other report

TASK 72

Analysis of defects

You are a member of a quality control team and your role is to inspect the quality of the finished product: a bolt.

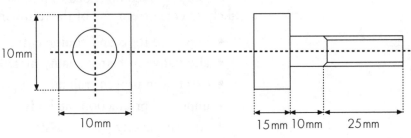

Bolt

There have been 120 defective products manufactured in the last month. Here is an analysis of those defects week-by-week.

Defect	Week no. 1	2	3	4	Total
Damaged thread	20	16	28	26	90
Long thread	4	2	1	3	10
Short thread	7	8	3	2	20
					120

Analysis of defects week-by-week

Here is another table; this shows on which day of the week the defective product was manufactured.

Day of the week defect occurred	Number of defects
Monday	42
Tuesday	12
Wednesday	14
Thursday	18
Friday	34
Total number	120

Analysis of defects by day of manufacture

Analyse the two tables and suggest potential causes of reduced quality and methods of preventing it.

APPENDIX

SUMMARY TABLES LINKING TASKS TO EVIDENCE REPORTS

The following tables show how the tasks within each chapter are put together to form the reports required for each element in a unit.

CHAPTER ONE

Section	Title	Task	Evidence
1.2	UK chemical manufacturing sector	1	Brief report
1.2	UK engineering manufacturing sector	2	Brief report
1.2	UK food, drink & tobacco manufacturing sector	3	Brief report
1.2	UK paper & board manufacturing sector	4	Brief report
1.2	UK printing & publishing manufacturing sector	5	Brief report
1.2	UK textiles, clothing & footwear manufacturing sector	6	Brief report
1.3	Reasons for the location of manufacturing companies	7	Brief report
		1–7	Report titled: 'Investigation of the importance of manufacturing to the UK economy', using the above evidence
1.4	Key production stages	8	Brief report & flow diagram
1.5	Scales of production	9	Brief report
1.6	Production systems	10	Flow diagram
1.7	Quality control	11	Flow diagram
1.8	Effects of changing the scales of production	12	Table
		8–12	Report titled: 'Investigate production systems', using the above evidence

Section	Title	Task	Evidence
1.9	Departmental functions in a manufacturing organization	13	Brief report
1.10	Departmental responsibilities during the manufacturing of products	14	Brief report & flow diagram
1.12	Work roles within departmental functions	15	Brief report & organization chart
1.13	Skill and training requirements for work roles	16	Brief report
		13–16	Report titled: 'Describe manufacturing organizations', using the above evidence
1.14	Local environmental impact of production processes	17	Brief report
1.15	Environmental impact of waste associated with production processes	18	Brief report
1.16	Environmental impact of different energy sources used in production processes	19	Brief report
1.17	Benefits of an organizational energy policy	20	Brief report
		17–20	Report titled: 'Identify the environmental effects of production processes', using the above evidence
2.3	Key design features	22	Record
2.4	Production process constraints	23	Record
2.5	Generating potential design proposals	24	Record
2.6	Assessing the feasibility of design proposals	25	Record
	Selecting your design proposal	26	Record
		22–26	Report titled: 'Originate product proposals from a given design brief', using the above evidence
2.7	Support materials for presentations	27	Record
2.8	Presentation techniques	28	Record
2.9	Technical vocabulary	29	Record
2.10	Presenting design proposals	30	Record
2.11	Feedback	31	Record
	Complete final product proposal	32	Record
		27–32	Report titled: 'Finalize proposals using feedback from presentations', using the above evidence

SECTION	TITLE	TASK	EVIDENCE
3.2	Product specifications	33	Lists
3.3	Key production stages	34	Descriptions and flow diagrams
3.4	Resource requirements	35	Tables
3.5	Processing times	36	Tables
3.6	Production schedules	37	Production schedule
3.7	Production plan	38	Production plan
		33–38	Report titled: 'Produce production plans', using the above evidence
3.8	Direct costs	39	Calculations
3.9	Indirect costs	40	Calculations
3.10	Total costs	41	Calculations
3.11	Effects of changing the scale of production on the costs of a product	42	Comments
		39–42	Report titled: 'Calculate the cost of a product', using the above evidence
3.12	Quality assurance systems	43	Brief report
3.13	Function of quality indicators & critical control points	44	Brief report
3.14	Role of testing & comparison in quality control	45	Brief report
3.15	Quality control techniques	46	Brief report
3.16	Use of test & comparison data	47	Brief report
3.17	Application of quality control	51	Flow diagram
		43–47 & 51	Report titled: 'Investigate quality assurance', using the above evidence
4.1	Key characteristics of materials	52	Lists
4.2	Processing methods	53	Brief report
4.3	Handling & storage of materials & finished products	54	Summary
4.4	Preparation of materials & components	55	Log
4.5	Preparation of equipment, machinery & tools	56	Log
4.6	Safety equipment, health & safety procedures & systems	57	Log
		52–57	Report titled: 'Prepare materials, components, equipment and machinery', using the above evidence

Section	Title	Task	Evidence
4.7	Safe operation of equipment, machinery & tools	58	Log
4.8	Control & adjustment of equipment & machinery	59	Log
4.10	Maintaining the level of materials to meet the production flow	60	Log
	Checking that safety equipment, health & safety procedures & systems are operational	61	Log
		58–61	Log titled: 'Process materials and components', using the above evidence
4.11	Equipment, machinery & tools used in the assembly process	62	Log
	Preparation & checking of materials & components used in the assembly process	63	Log
4.13/ 4.14	Assembly of materials & components to specification & quality standards	64	Log
4.15/ 4.16	Finishing methods	65	Log
	Correct use of safety equipment, health & safety procedures & systems during assembly & finishing	66	Log
		62–66	Log titled: 'Assemble & finish materials & components to specification', using the above evidence
4.17	Quality indicators & frequency of analysis	67	Flow diagrams
4.18	Test & compare samples at critical control points	68	Flow diagrams
4.19	Formats for recording data	69	Tables and graphs
4.20	Defects & their causes	70	Brief report
4.21	Dealing with defective products	71	Brief report
		67–71	Report titled: 'Apply quality assurance to manufactured products', using the above evidence

INDEX